Environmental and Ecological Statistics

Environmental and Ecological Statistics

Miles Wilson

MURPHY & MOORE
www.murphy-moorepublishing.com

Environmental and Ecological Statistics
Miles Wilson
ISBN: 978-1-63987-193-3 (Hardback)

MURPHY & MOORE

Published by Murphy & Moore Publishing,
1 Rockefeller Plaza,
New York City, NY 10020, USA

Cataloging-in-Publication Data

Environmental and ecological statistics / Miles Wilson.
p. cm.
Includes bibliographical references and index.
ISBN 978-1-63987-193-3
1. Environmental sciences--Statistical methods. 2. Environmental sciences. 3. Statistics. I. Wilson, Miles.
GE45.S73 S73 2022
363.700 727--dc23

Table of Contents

Preface

The application of methods from the field of statistics to study the undisturbed state of the environment, interactions between humans and their environment, and the urban environment is known as environmental statistics. The discipline often overlaps with the fields of social and economic statistics. Environmental statistics finds extensive application across various fields such as risk regulation and control, pollution, meteorology, public health and safety, etc. Some of the different types of studies covered under this discipline to detect environmental changes and accidents are baseline studies, targeted studies and regular monitoring. The topics included in this book on environmental statistics are of utmost significance and bound to provide incredible insights to readers. It aims to shed light on some of the unexplored aspects of environmental statistics. This book will serve as a valuable source of reference for those interested in this field.

A detailed account of the significant topics covered in this book is provided below:

Chapter 1- The branch of study which deals with the collection, organization, analysis, presentation and interpretation of data is known as statistics. It can be broadly categorized into qualitative and quantitative data which can be further classified into nominal, ordinal, interval and ratio data. This chapter provides an introduction to statistics as well as these types of data.

Chapter 2- The branch of mathematics which deals with the likelihood of the occurrence of an event is known as probability. A few of the different types of events which are studied within probability are simple event, compound event, exhaustive event and dependent event. All the diverse aspects of these types of events in probability have been carefully analyzed in this chapter.

Chapter 3- The mathematical function which provides the probabilities of the occurrence of different outcomes for a particular trial is known as probability distribution. The topics elaborated in this chapter will help in gaining a better perspective about the major concepts studied in relation to probability distributions such as probability functions of discrete random variables and continuous probability distributions.

Chapter 4- Determinants and matrices are used in linear algebra to non-homogenous equations in linear form. Determinants can only be used for square matrices. This chapter has been carefully written to provide an easy understanding of the varied facets of determinants and matrices such as special matrices, adjoint and inverse of a matrix, and rank of a matrix.

Chapter 5- The design of any task which seeks to describe and explain the variation of information within conditions which are hypothesized to reflect the variation is known as experimental design. This chapter closely examines the key concepts of experimental design, such as completely randomized design and randomized block & Latin square design, to provide an extensive understanding of the subject.

Chapter 6- The series of data point which are indexed in time order is known as time series. A few of the major concepts studied in relation to time series are modelling of time series, stationary time series and non-stationary time series. The topics elaborated in this chapter will help in gaining a better perspective about these aspects of time series.

It gives me an immense pleasure to thank our entire team for their efforts. Finally in the end, I would like to thank my family and colleagues who have been a great source of inspiration and support.

Miles Wilson

Basics of Statistics

The branch of study which deals with the collection, organization, analysis, presentation and interpretation of data is known as statistics. It can be broadly categorized into qualitative and quantitative data which can be further classified into nominal, ordinal, interval and ratio data. This chapter provides an introduction to statistics as well as these types of data.

Statistics is a science that uses different tools and techniques to organize the data in the descriptive form (in the form of table, graphs and pictorial presentation) and extract information from the data that helps in making decisions.

Statistics is considered as a study of collecting, analyzing, interpreting, presenting and organizing the data. Statistics consists of a body of methods for collecting and analyzing data.

Statistics is basically used to get insight into the data. Statistics helps to get an overall picture of an objective from the data based on graphical presentation or numerical calculation without any constraints on the number of observations. It not only gives graphical representation but also makes inference and predict relations among variables.

For example, opinion polls for election involve designing of questionnaire then collection of data from the field with appropriate sampling methods. Further collected data is analyzed using statistical methods. At the last, the results are present in a understandable manner.

Classification of Statistics

Statistical methods have different branches and each branch has its great importance in the literature as well as for practical point of view. The following model explains the relation between these methods:

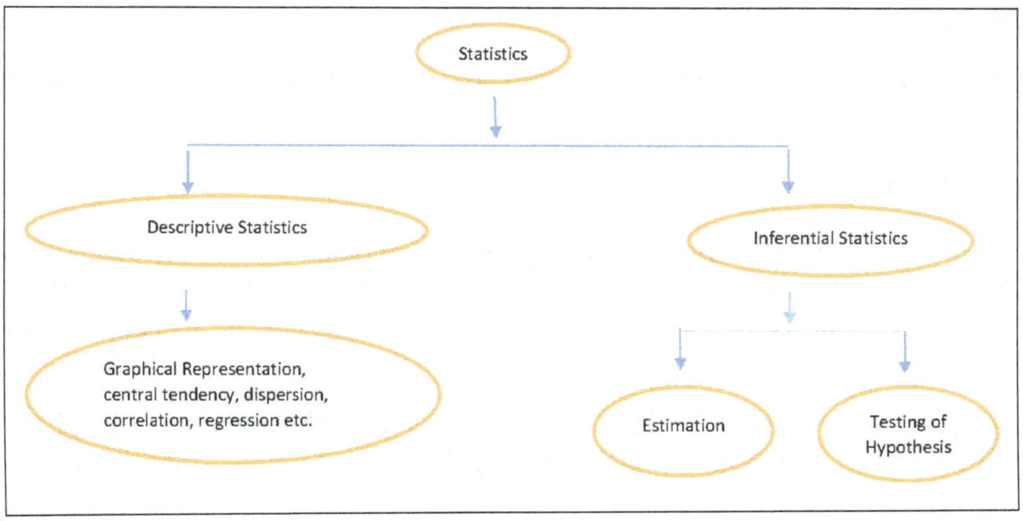

From the above model, one can see that for statistical methods are segmented into two methods. One is descriptive statistics and second is inferential statistics.

Descriptive Statistics

In Descriptive Statistics, graphical representation methods like histogram, bar chart, pie chart, ogive curves, box plot, line chart etc. are used to present data in graphical form. Descriptive methods also deal with the summary of the data. This includes these following measures:

Central Tendency Measures

For example mean, median, mode.

Measure of Dispersion

Variance, standard deviation, range.

There are other measures like skewness, kurtosis, quartiles, quantile, percentile etc. These measures are used to study other characteristics of the data. In descriptive statistics, one can study the scatterplot, correlation between variables, regression analysis.

Basically descriptive statistics deals with quantitatively describing the features of a data available from the sample/population.

Inferential Statistics

Descriptive statistics is used just to represent the information available in the data. But sometimes it is difficult to collect all the observation in the study for analysis. In this situation, sampling technique is used to collect element of the population in a non-random/random manner. Then inferential statistics is used to extract information from the sample as well to test the significance of an objective based on observations using testing of hypotheses. It is further segmented into two parts. These are:

- Estimation: Estimation techniques is to extract information about the population from the sample.
- Testing of Hypotheses: This method is used to test the significance of a statement about a problem.

How Statistics Works?

In a very simple language, statistics is basically used to find solution for an objective or a problem in different areas for e.g. if a person is interested in calculating the height of the students of college in particular region of a state for study. Now the question is how to approach this problem.

The first thought that come to your mind is how we can say anything about the height among college students as there is variations in heights among the students. Now, what to do?

By using statistical methods, tools and techniques, one can find a numerical value that can help a person to find out solution. In this case, one can use very simple method that is measures of central tendency for e.g. mean, median, mode. "Mean" is generally used method in our daily life and is considered as the average of given values of a thing that has same characteristics. So take a mean of the values and see the outcome. The outcome will depend on the observation that we have. If the observations have little variation in term of height then average (mean) value is very close to observations values. If the observations are not homogenous (similar) in terms of height then the average value is different from observations values.

For e.g.: The values of the height of the 8 students in feet are 5' 2", 5' 4", 5' 6", 5' 3", 5' 6", 5' 3", 5' 5", 5' 6" and their mean value* is 5'4" (approx).

If the values of the height of the 8 students are 4' 3", 4' 9", 5' 9", 4'1", 5' 1", 4' 8", 4' 5",5' 3" and their mean value is 4' 8" (approx).

Now from the above example, one can see that there is major difference between the mean values of the two data.

Need of Statistics

We will discuss the need of statistics in our life. We are using statistics or basic mathematics intentionally or unintentionally in our daily life. There are many examples like while buying any item from the market people inspect a smallest fraction of an item to check the quality. In statistics, we call this selected item as "sample" and whole lot of an item from which item is selected called as "population".

Nowadays, we are more dependent on information technology that enables us to develop different tools and devices for different purposes. For example, mobile phone manufacturer companies want to increase the credibility of their product. That will increase the branding of their company's name. For these reasons, companies spend a lot on the Research and Development department. To test the reliability of products statistical tools and techniques based on data are used. But with increase in the complexity of the system, new tools and techniques are required which are more efficient and less costly. To fulfill the demand of data analysis consistent research is going on to cater this need. Hence, there is consistent increase in the demand of statistics tools and techniques.

We can see the descriptive presentation of data in day to day life for e.g.one can see the charts and graphs of inflation, opinion polls, economic growth etc on newspapers, television and other communication resources. Figure represents the trend of inflation per year from the year 2001-2006. In this figure, one can observe that in 2002 inflation was less than in 2001.There was an increasing trend of inflation from 2001- 2004 and it declined till 2006.

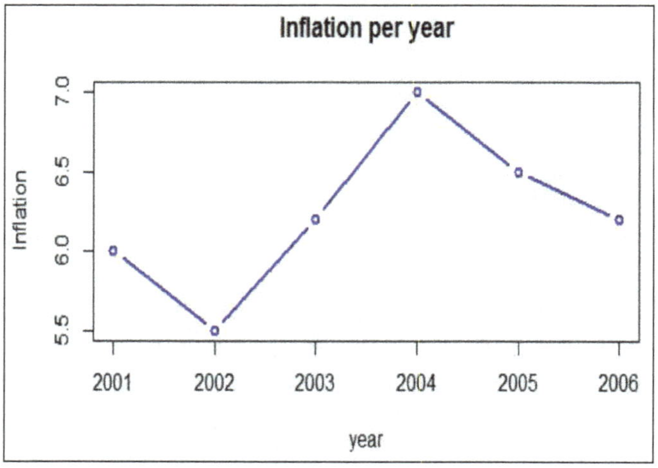

The purpose of graphical representation helps in (descriptive presentation helps) understanding the objective in a better way. It is easy for a layman person to understand things.

For example, if one has to study the average expenses of a middle class family on food in a particular small town. One can choose using sampling technique some houses from the list of available houses then collect responses from them. This data is analyzed further to get an estimate of the average expenses on food. One cannot determine the average expenditure on meal of a middle person based on single value. Statistics methods also help to give an interval containing lower and upper values. So, one can predict that expenses on food lie between this interval.

Importance of Statistics

As Statistics has a great role in different areas like determining the existing status of per capita income, mortality rate, inflation, population density rate etc. This is the reason that statistics holds a great importance in almost every field like Industry, Economics, Mathematics, Astronomy and Nursing.

In Industry

In Industry, statistics has its great importance. Industry growth depends on the demand and supply of goods, customer loyalty and market stability. Statistics helps businessman to plan production by taking into account all the possible factors that effects the sale of goods. Quality of products can also be checked by using statistical methods. Hence statistics has great importance in the working of an Industry.

In Economics

Economics deals with the study of factors that affects the economy of the world. So basically economics depends upon the figures of these factors and with passage of time these figures keep on changing.

Economists use statistical tools and techniques to predict these factors with accuracy. In economics, statistical techniques are used for the collection of the data and as well as for analysis purposes. Statistics is also used to compare the factors like demand and supply, import and export, inflation rate, the per capital income etc. Economists use descriptive statistics tools like presenting data in tabular form, histogram, pie charts etc to present their objectives in more attractive as well easily understandable way.

In Mathematics

Statistics and Mathematics are complementary subjects. In mathematics, we use descriptive statistics as well as inferential statistics like central tendency measures, dispersion measures and estimation, hypothesis testing etc. In statistics, we use integration, derivative and basic algebra for analysis purposes. As mathematical tools and techniques can only give you most reliable results if they are based on strong evidence. Statistical methods are used for collecting these evidences with higher probability of more precisely. For example, to check the viability of a project introduced by a large firm is calculates using both mathematics and statistics techniques. Hence, mathematics and statistics methods are both enrooted with each other. Statistics is also called branch of applied mathematics.

In Banking

Statistics has a great importance in banking field. As banks principle duty is to deposit money of depositors and use that money for issuing loans to lenders. There are many factors like rate of interest, economy stability etc. that determine the business of the bank. These factors keep on varying with time. Hence it is very important for banks to predict these factors more precisely as possible. Statistics techniques play an important role in predicting these factors with higher probability. Also statistics helps bank to predict defaulters of loans. For example, if the market will crash then people will lose their confidence in the economy of the country as well as bank. Also lenders will default on their loan amount due to lack of demand of their goods that will decrease their profit. Statistics can lower the bank's risk by predicting factors that effects the market.

In Administration

Statistical methods are basically used for collection as well as analysis of data and also for the interpretation of the results in different sectors of administration. In government offices, most of the projects for social welfare depend on it successful working. For example, if the government wants to increase the dearness allowance or revise the pay scale of employees due to increase in the living cost. Statistical methods will be used to determine the rate of inflation that affects the cost of living. Also central and state government budgets depends upon the statistical tools that help them in estimating income and expenditure from different sources. So statistics has a great importance in the administration's work.

In Accounting and Auditing

Accounting is used by firms to maintain the balance sheets. It is based on data and it does not survive without the authenticity of the data. In accounting, statistics is used for decision making purposes for example how to correctly quantify the assets of the company, liquidity value of the assets, how to calculate depreciation cost of the assets etc. All these decisions are based on the market conditions. For example, all electronic assets value depleted within months due to high new technology development rate.

In Natural and Social Sciences

Statistics has its vital role in natural and social sciences. In natural sciences, for research and development purpose laboratory experiments and field experiments are conducted. Information is gathered in the form of the data. Now to draw conclusion from the data statistical methods are used. For example, a scientists are interested to test the effectiveness of newly developed drug for the heart patients. To test the effectiveness, they collect the information of health status of heart patient after taking dose of new medicine. It is possible that drug is effective for some patients and ineffective for others. So, statistical method especially hypothesis testing is used to test the significance of newly developed drug for heart patients with some confidence.

In Social Sciences, most of the research work is done with the help of questionnaire and field work. Statistics is used to test the reliability and validity of the questionnaire. It is used to analyze the data obtained from the questionnaire and draw inference from them. For example, most of the social workers collect information through questionnaire. These questionnaire are further processed and stored in the form of data. This data is analyzed further and draw conclusions. In field work, statistics also help social scientists to determine the sample size and to adopt which sampling techniques.

In Astronomy

Astronomy deals with the densities of heavenly bodies, masses, measurement of distance between stars and planets, their sizes as well as other activities in the space. If errors occur while calculation of these figures that loss is unbearable for all. So, to avoid such type of errors statistical methods are used. For example, from the old days, least squares method is used to find the movements of stars.

In Demography

Statistics is considered as a backbone of demography. Demography deals with the study of population structure, sex ratio, health status and age group among population. It is difficult to collect information from each individual. Statistical methods are used to determine sample size and sampling technique based on objective. For example, if a person is interested in studying the mortality rate of cancer patients in a state. For collecting data, person must have to determine the resources from where one can collect information. After that appropriate sampling technique is used to collect samples. Based on these values one can estimate the mortality rate of cancer patients in a state.

Limitations of Statistics

- Statistical analysis are mostly used for quantitative data. However, there are many tools available that can be used for qualitative analysis.
- It can only be applied collective and not on single events.
- Statistics give results on average basis that are true in the long run and not able to give result for a particular case.

- Statistical results are not percent correct. These are just approximate values and not exact one.
- Statistics approach may not reveal the entire problem as some of the factors are incapable of statistical analysis.
- Statistical methods may lead to wrong decision if not applied in a correct manner.
- Statistics tools, if applied in a wrong manner, can lead to major loss.

Types of Data

When it all boils down to it, all data that you collect are either measured or are an observed feature of interest, and at the highest level that gives us 2 kinds of data:

- Quantitative data.
- Qualitative data.

Quantitative data is information about quantities of things, things that we measure, and so we describe them in terms of numbers. As such, quantitative data are also called Numerical data.

On the other hand, qualitative data give us information about the qualities of things. They are observed phenomenon, not measured, and so we generally label them with names. Qualitative data are also known as Categorical data.

2 Types of Quantitative Data - Discrete Data and Continuous Data

There are 2 types of quantitative data – Discrete data and Continuous data.

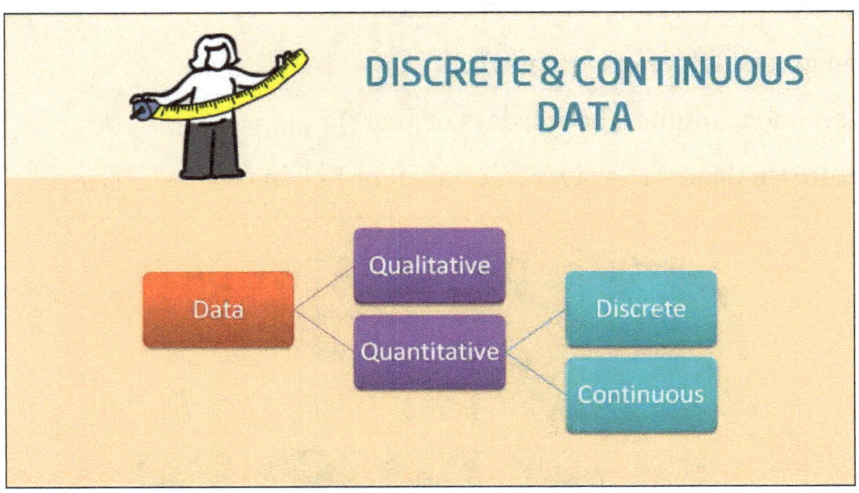

Discrete data is information that can only take certain values and can't be made more precise. This might only be whole numbers, like the numbers on a die (any number from 1 to 6) or could be other types of fixed number scheme, such as shoe sizes (2, 2.5, 3, 3.5, etc.). They are called discrete data because they have fixed points and measures inbetween do not exist (you can't get 2.5 on a die, nor can you have a shoe size of 3.49).

Counted data are also discrete data, so numbers of students in a class, number of patients in the hospital and number of marbles in a bag are all examples of discrete data.

On the other hand, continuous data is data that can take any value, usually within certain limits, and could be divided into finer and finer parts. A person's height is continuous data as it can be measured in metres and fractions of metres (centimetres, millimetres, nanometres). Time of an event is also continuous data and can be measures in years and divided into smaller fractions, depending on how accurately you wish to record it (months, days, hours, minutes, seconds, etc.).

Examples of Quantitative Data

Quantitative data are measured with some kind of measuring implement - ruler, jug, weighing scales, stop-watch, thermometer and so on.

Here are some examples of quantitative data that can be measured with a ruler or measuring tape:

- Height (e.g. 1.65 metres).
- Distance (e.g. 1.47 miles).
- Waist circumference (e.g. 37 inches).

Examples of quantitative data that you can measure with a jug include:

- Water (e.g. 1.7 litres).
- Flour (e.g. 400 millilitres).
- Whisky shots (1.5 fluid ounces).

Other things that are examples of quantitative data are:

- Weight (in grams, kilograms, tonnes).
- Time (in seconds, minutes, hours, days or years).
- Temperature (in degrees Celsius, Fahrenheit or Kelvins).

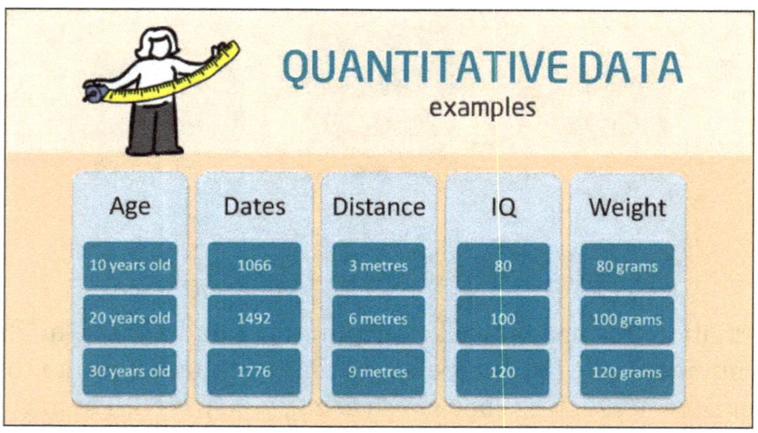

Examples of Qualitative Data

Qualitative data are observed data that are placed into categories. This could be anything from the colour of eyes, country of origin, animal species, to emotional state, level of opinion, political views and religious beliefs.

Here are some examples of qualitative data:

- Socioeconomic status (Lower class, Middle class, Upper class).
- Opinion (Agree, Neutral, Disagree).
- Hair colour (Red, Brown, Blonde).

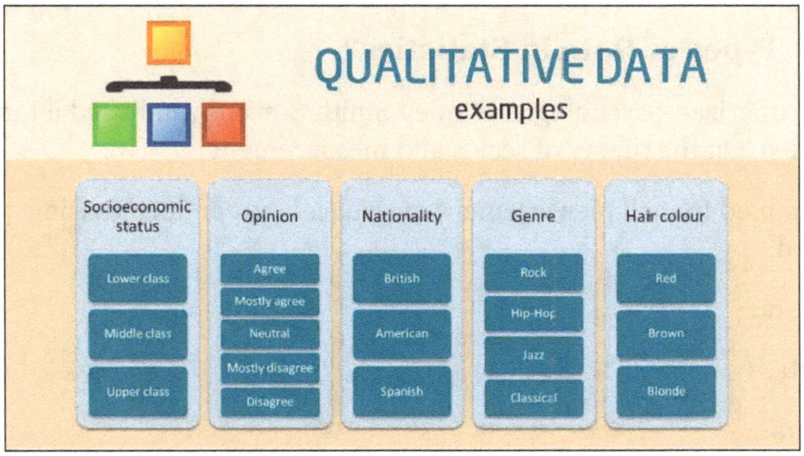

How can you Tell if Data are Quantitative or Qualitative?

What is the difference between quantitative data and qualitative data? Simply put:

- Quantitative data are measured.
- Qualitative data are observed and placed into categories.

In practical terms, when data are text-based, such as Types of Animals [Pig, Sheep, Cow], then they are qualitative data. When they are number-based, like Length (2.13 metres or 5.72 miles) they are quantitative data.

The one exception to this is when categories have been numbered for practical purposes, such as Types of Animals [1, 2, 3] instead of [Pig, Sheep, Cow]. In this case, the numbers must be treated as the names of the categories - you're not allowed to do any calculations with them.

Qualitative data are categorised.

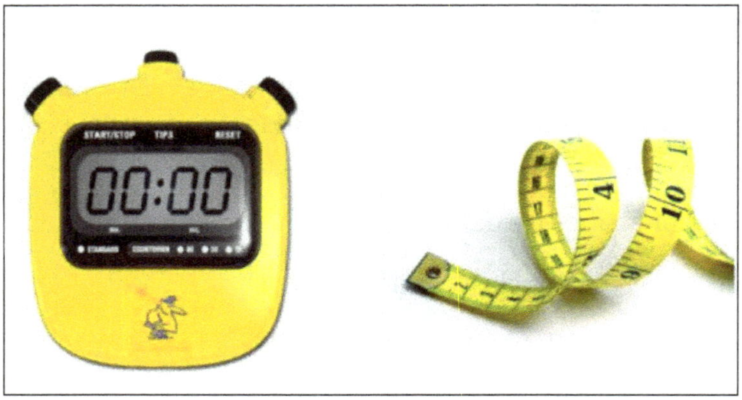

Tools for measuring quantitative data.

What are the 4 Types of Data in Statistics?

Back in 1946, the American psychologist Stanley Smith Stevens published a famous paper in the journal Science titled 'On the theory of scales and measurement'.

In this paper he claimed that all measurement in science was conducted using 4 different types of scales that he called:

- Nominal data.
- Ordinal data.
- Interval data.
- Ratio data.

This paper essentially unified Qualitative data (Nominal data and Ordinal data) and Quantitative data (Interval data and Ratio data).

Quantitative data and Qualitative data can be sub-divided into 4 further classes of statistical data types; Ratio, Interval, Ordinal and Nominal Data, like this:

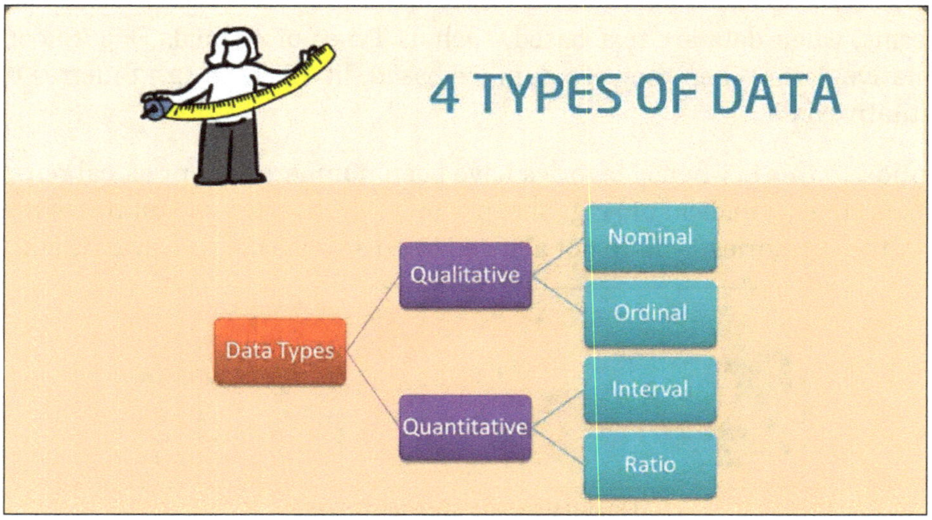

You can figure the difference by asking just 2 questions:

- Quantitative data or Qualitative data - Have the data been measured (quantitative data) or were they observed (qualitative data)?
- (If quantitative): Meaningful Zero - Does the scale of measurement include a unique, non-arbitrary zero value? (answer: Yes = Ratio data, No = Interval data), else.
- (If qualitative): Order - Can some sort of progress be detected between adjacent data points or categories? Can the data be ordered meaningfully? (answer: Yes = Ordinal data, No = Nominal data).

It's useful for you to remember that if you can answer just these 2 questions then you can correctly determine each of your 4 types of statistical data.

And if you're wondering about the relationship between Discrete Data and Continuous Data, and each of Ratio and Interval data, either of Discrete data and Continuous data can be Ratio data or Interval data.

If you have identified that your variable is Discrete data, then remind yourself of the questions above and ask yourself whether your data have a true zero. If they do, they are Ratio data, and if not, they are Interval data.

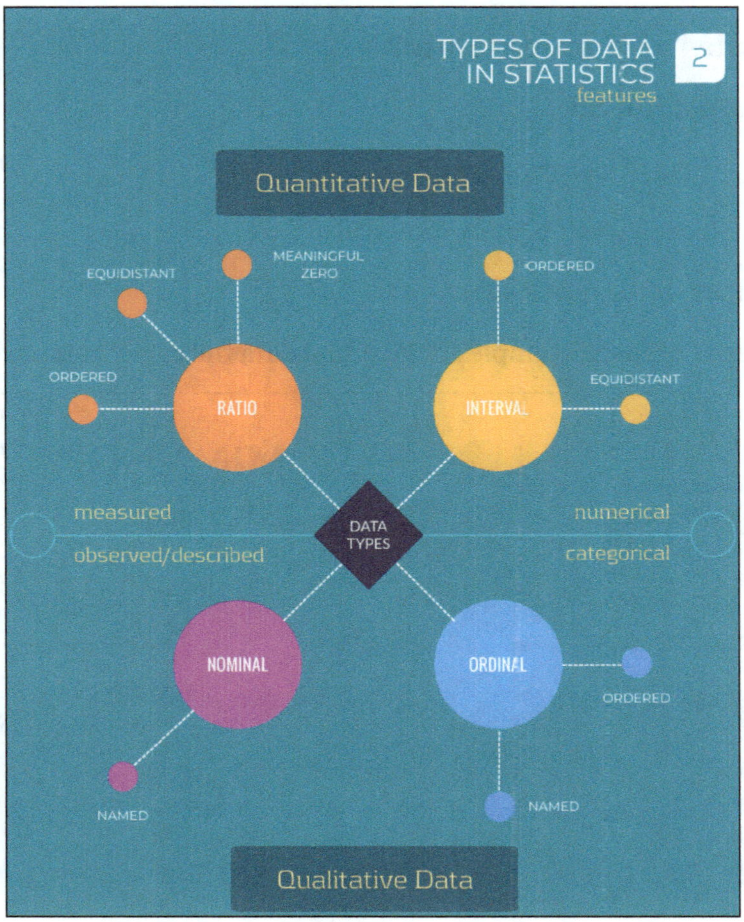

What type of Data is Nominal Data?

Nominal data is the statistical data type that has the following characteristics:

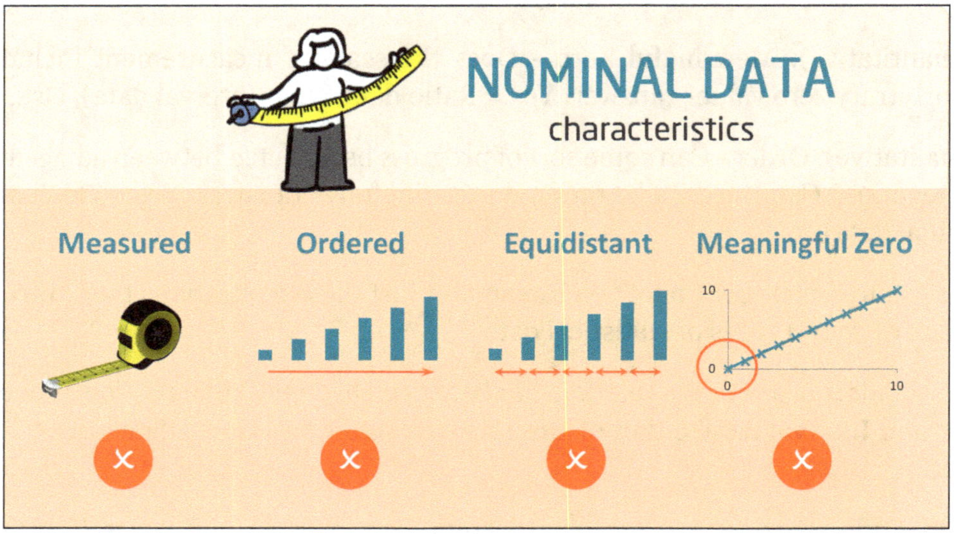

Nominal Data are observed, not measured, are unordered, non-equidistant and have no meaningful zero.

We can differentiate between categories based only on their names, hence the title 'nominal'.

What are Examples of Nominal Data?

Examples of Nominal data include:

- Nationality (British, American, Spanish).
- Genre/Style (Rock, Hip-Hop, Jazz, Classical).
- Favourite colour (red, green, blue).

You can see that in each of these examples of Nominal data the categories have no order.

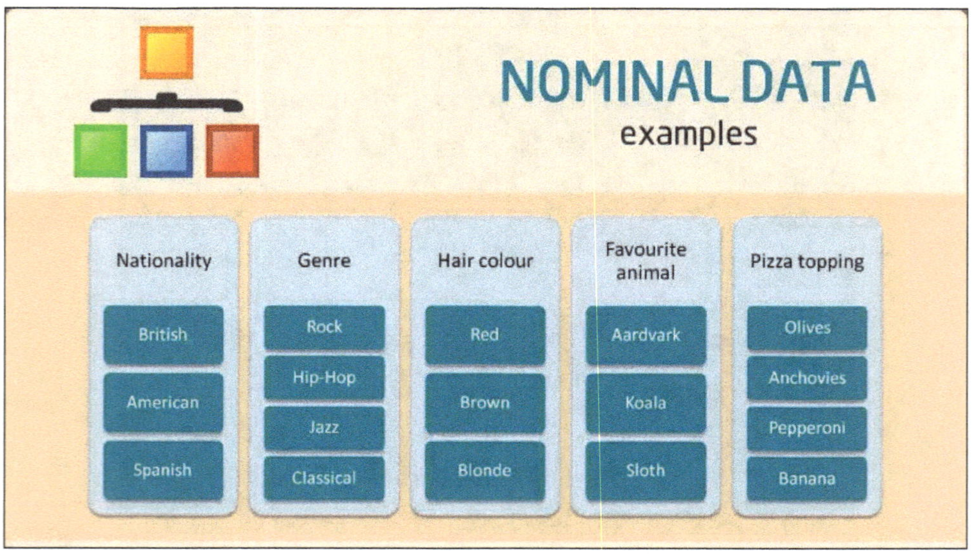

What can you Calculate with Nominal Data?

The only mathematical or logical operations you can perform on Nominal data is to say that an observation is (or is not) the same as another (equality or inequality), and you can use this information to group them together.

You can't order Nominal data, so you can't sort them. Neither can you do any mathematical operations because they are reserved for numerical data.

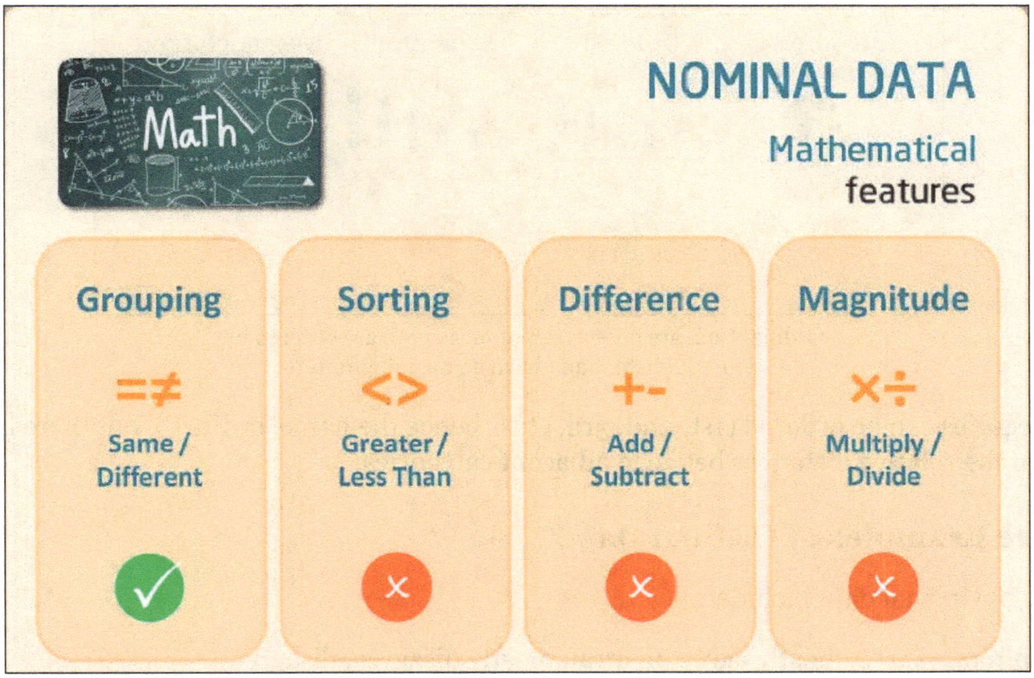

What Descriptive Statistics can you do with Nominal Data?

With Nominal data you can calculate the following:

- Frequencies - Count how many you have in each category.
- Proportions - Determine how often something happens by dividing the frequency by the total number of events.
- Percentages - Transform the proportions to percentages by multiplying by 100.
- Central point - You can determine the most common item by finding the mode.

Other ways of finding the middle of the class, such as median or mean make no sense because ranking is meaningless for nominal data.

What Else can you do with Nominal Data?

If you need to know more about Nominal data and what you can do with them, including descriptive statistics, data visualisations, creating dummy variables and which statistical tests you can use with them.

What Type of Data is Ordinal Data?

Ordinal data is the statistical data type that has the following characteristics:

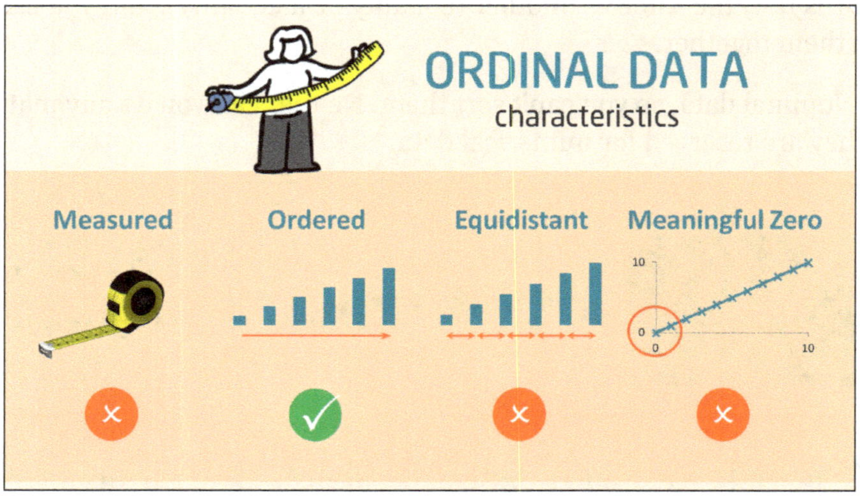

Ordinal Data are observed, not measured, are ordered but non-equidistant and have no meaningful zero.

Their categories can be ordered (1st, 2nd, 3rd, etc. - hence the name 'ordinal'), but there is no consistency in the relative distances between adjacent categories.

What are Examples of Ordinal Data?

Examples of Ordinal data include:

- Opinion (agree, mostly agree, neutral, mostly disagree, disagree).
- Tumour Grade (1, 2, 3).
- Time of day (morning, noon, night).

In the examples of Ordinal data above, there is an obvious order to the categories.

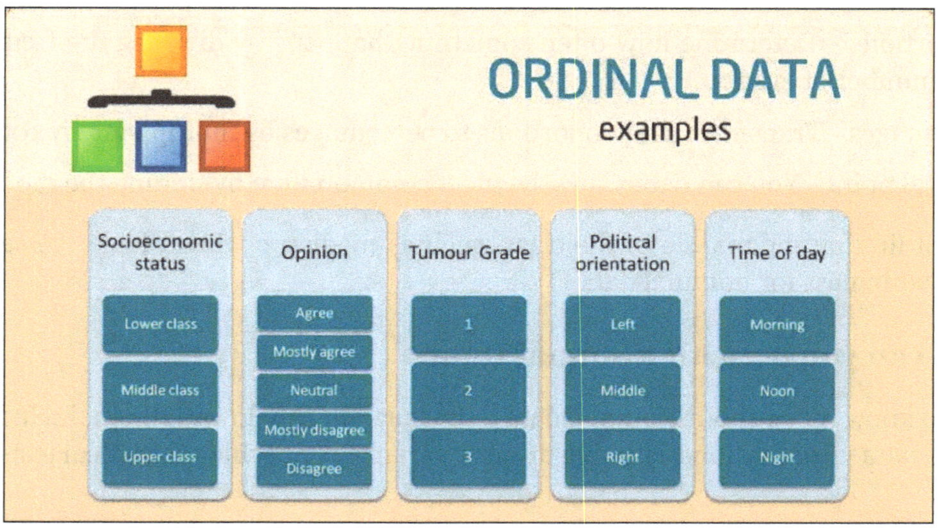

What can you Calculate with Ordinal Data?

With Ordinal data, as with Nominal data, you can group the data by assessing whether they are the same or different.

As Ordinal data are ordered, they can be sorted by making simple comparisons between the categories, such as Greater/Less than, More/Less, Higher/Lower, etc.

You can't do any mathematical operations with Ordinal data, though, because they aren't numerical data.

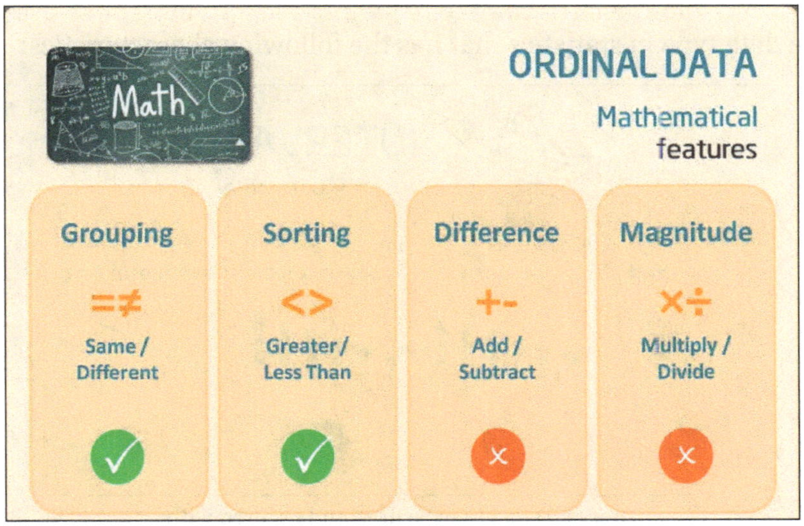

What Descriptive Statistics can you do with Ordinal Data?

With Ordinal data you can calculate precisely the same things as you can with Nominal data, with a couple of extra things. That is:

- Frequencies - Count how many you have in each category.
- Proportions - Determine how often something happens by dividing the frequency by the total number of events.
- Percentages - Transform the proportions to percentages by multiplying by 100.
- Central point - Since there is an order to the data you can rank them and compute the median (or mode, but not the mean) to find the central value.
- Summary statistics - As the data are ordered, you can use percentiles and the inter-quartile range to summarise your data.

What Else can you do with Ordinal Data?

If you need to know more about Ordinal data and what you can do with them, including descriptive statistics, data visualisations, creating dummy variables and which statistical tests you can use with them.

What is the difference between Ordinal Data and Nominal Data?

Ordinal data and Nominal data are both qualitative data, and the difference between them is that Nominal data can only be classified - arranged into classes or categories - whereas Ordinal data can be classified and ordered.

One of the assumptions of Ordinal data is that although the categories are ordered, they do not have equal intervals.

What Type of Data is Interval Data?

Interval data is the data type in statistics that has the following characteristics:

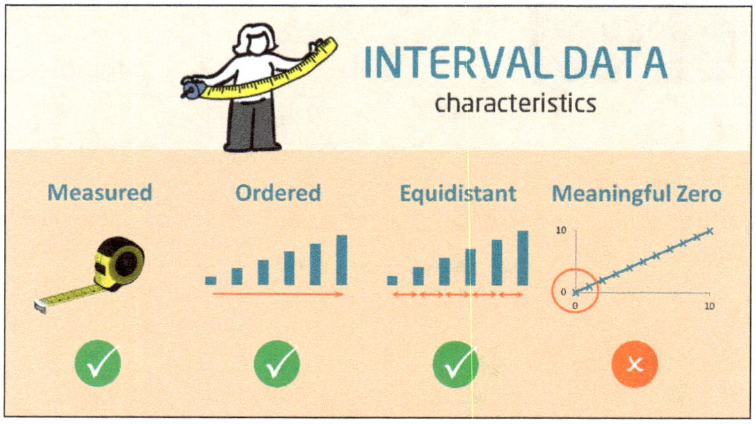

Interval Data are measured and ordered with equidistant items, but have no meaningful zero.

Interval data are ordered, can be continuous (have an infinite number of steps) or discrete (organised into categories), and the degree of difference between items is meaningful (their intervals are equal), but not their ratio.

The key points of an Interval scale is that the word 'interval' means 'space inbetween', which is the important thing to remember - interval scales not only tell us about order, but also about the value between each item.

What are Examples of Interval Data?

Examples of Interval data include:

- Temperature (°C or F, but not Kelvin).
- Dates (1066, 1492, 1776, etc.)
- Time interval on a 12 hour clock (6am, 6pm).

Crucially, Interval data can be negative, whereas Ratio data cannot.

Although Interval data can appear very similar to ratio data, the difference is in their defined zero-points. If the zero-point of the scale has been chosen arbitrarily (such as the melting point of water or from an arbitrary epoch such as AD) then the data cannot be Ratio data and must be Interval data.

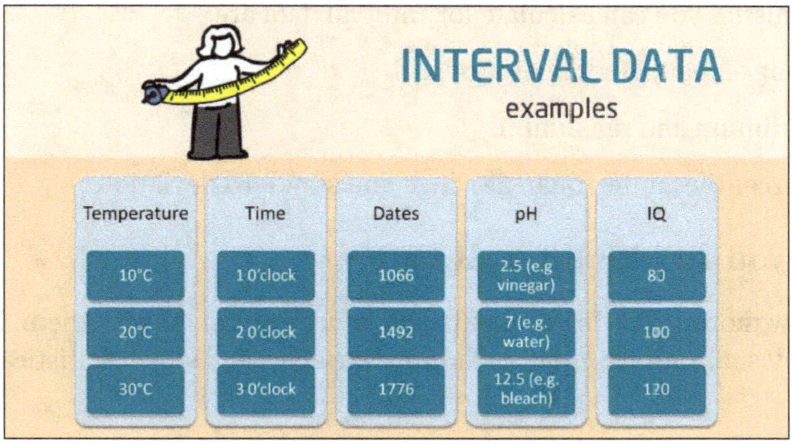

What can you Calculate with Interval Data?

With Interval data you can compare the degrees of the data (equality/inequality, more/less) and you can also add/subtract the values.

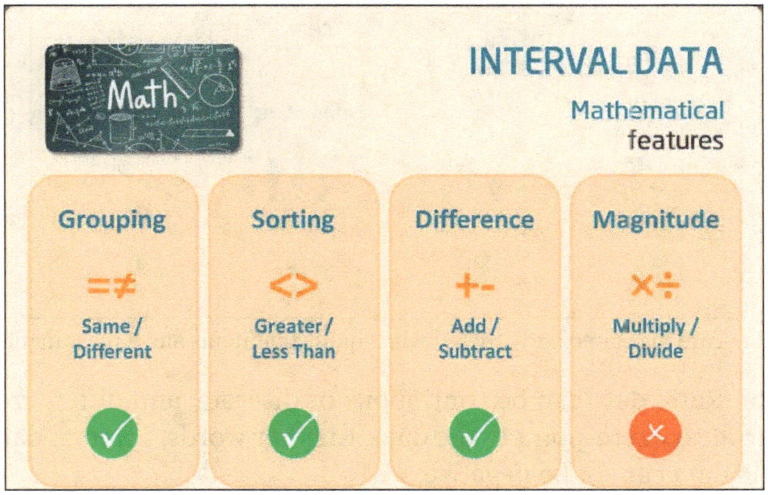

For example, with Interval data you can say things such as '28 °C is 8 degrees hotter than 20 °C' (28 - 8 = 20) or '6pm is 4 hours later than 2pm' (2 + 4 = 6).

However, you cannot multiply or divide the numbers because of the arbitrary zero, so you can't say 'a person with an IQ of 150 is 3 times as smart as a person with an IQ of 50' or '6pm is twice as late as 3pm'.

What Descriptive Statistics can you do with Interval Data?

As Interval data are quantitative data (continuous data), the descriptive statistics for these are very different to those for qualitative data.

The central value of Interval data is typically the mean (but could be the median or mode). You can also express the spread or variability of the data using measures such as the range, standard deviation, variance and/or confidence intervals.

The descriptive statistics you can calculate for Interval data are:

- Central Point - Mean, Median or Mode.
- Range - Minimum and maximum.
- Spread - Percentiles, inter-quartile range and standard deviation.

What Else can you do with Interval Data?

If you need to know more about Interval data and what you can do with them, including descriptive statistics, data visualisations, creating dummy variables and which statistical tests you can use with them.

What Type of Data is Ratio Data?

Ratio data is the data type in statistics that has the following characteristics:

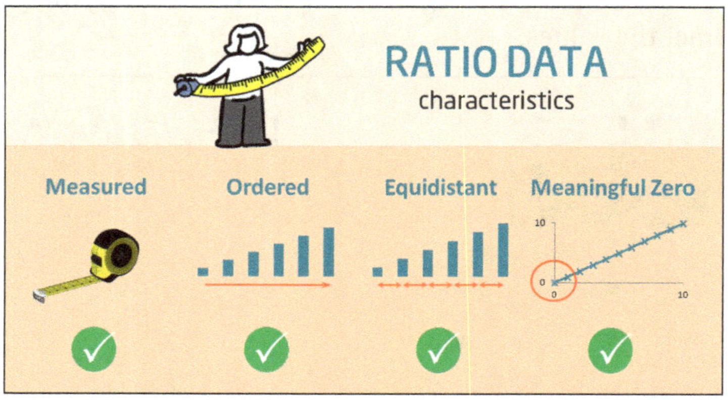

Ratio Data are measured and ordered with equidistant items and a meaningful zero.

As with Interval data, Ratio data can be continuous or discrete, and differs from Interval data in that there is a non-arbitrary zero-point to the data. In other words, Interval data can take negative values, whereas Ratio data cannot be negative.

What are Examples of Ratio Data?

Examples of Ratio data include:

- Age (from 0 years to 100+).
- Temperature (in Kelvin, but not °C or F).
- Distance (measured with a ruler or other such measuring device).
- Time interval (measured with a stop-watch or similar).

For each of these examples of Ratio data there is a real, meaningful zero-point. The age of a person, absolute zero, distance measured from a pre-determined point or time all have real zeros.

With real zero-points we can multiply and divide the numbers - a 6 year old is half the age of a 12 year old and matter at 100K has half the energy of matter at 200K. Similarly, the distance from

Barcelona to Berlin is half the distance as Barcelona to Moscow, and it takes me twice as long to run the 100m as Usain Bolt but only half the time of my Grandad.

In the examples of Ratio data above, can you see why Temperature measured in °C or F are examples of Interval data, whereas Temperature measured in Kelvins is an example of Ratio data?

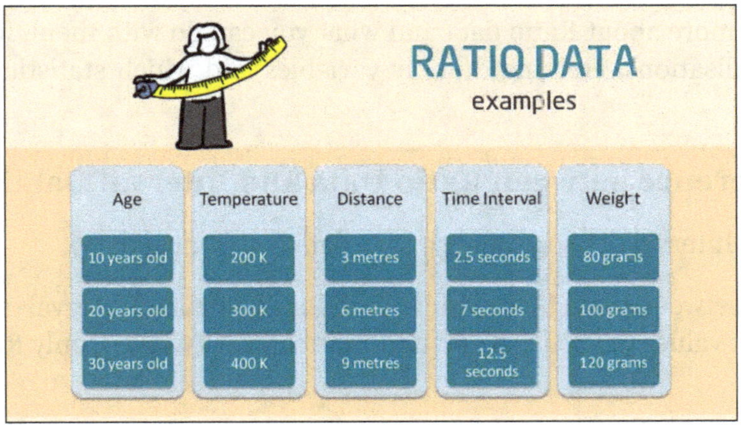

What can you Calculate with Ratio Data?

With Ratio data, all cards are on the table. You can compare the data (equal or not), you can sort the values (greater/less than), add/subtract them, and you can also multiply or divide them.

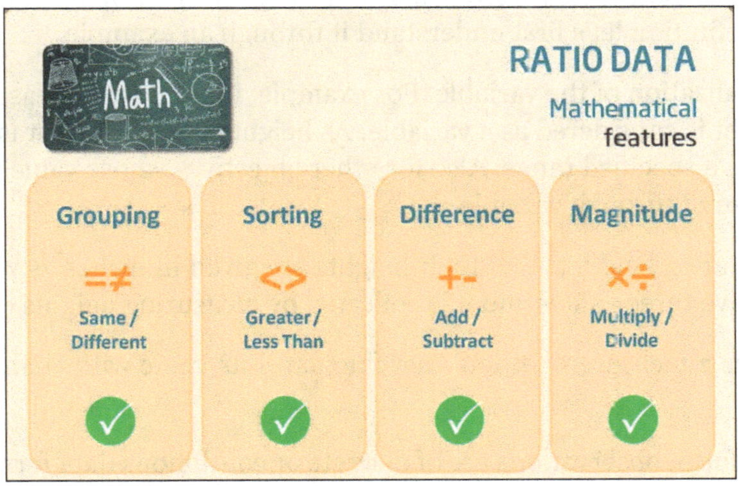

For example, with Ratio data you can check whether your apples weigh the same, you can sort them from heaviest to lightest and you can calculate the differences in their weights.

With Ratio data you can also multiply and divide your numerical data, so you can calculate the ratios of their weights and say things like 'this 100 gram apple weighs twice as much as this 50 gram apple'.

What Descriptive Statistics can you do with Ratio Data?

With Ratio data you can calculate precisely the same things as you can with Interval data. That is:

- Central Point - Mean, Median or Mode.

- Range - Minimum and maximum.
- Spread - Percentiles, inter-quartile range and standard deviation.

What Else can you do with Ratio Data?

If you need to know more about Ratio data and what you can do with them, including descriptive statistics, data visualisations, creating dummy variables and which statistical tests you can use with them.

What is the difference between Ratio Data and Interval Data?

Both Ratio data and Interval data are quantitative data (numerical data).

The only difference between them is that while both Ratio data and Interval data have equal spacing between adjacent values (so you can add and subtract their values), only Ratio data have a true zero.

That is, Interval data can take negative values, whereas Ratio data cannot - and this means that Ratio data can be multiplied and divided, but Interval data cannot.

Variable

Variable is another commonly used word when collecting information from the observations. Before looking at the definition let's first understand it through an example.

Basically data are realization of the variable. For example, in a study of measuring average height of student then height is considered as a variable. As height of each student is different and it can take any value within a specified range. Also for other objectives whose value vary under different conditions are measured through variable.

For example it there are 3 students and their heights are given in inches as 78, 81 and 56 respectively. So our data have three values and it is collected by measuring heights of students.

Variable is defined as a measurement tool that fluctuates its value with change in the conditions of an objective.

In quantitative data form, observations are of discrete or continuous data form. Hence there must be two variables. These are:

- Discrete variables.
- Continuous variables.

Discrete Variables

As variable is considered as an instrument that can change its values. Discrete variables are those variables that can only change values in whole numbers. For example, numbers of boys in a class, number of patients in a hospital, number of accidents occurred in a year and number of states in a country.

Variables that can only take on a finite number of values is termed as discrete variables. This value vary from one observation to the other.

Continuous Variables

Height of different person vary and also it can take values in decimals so height is considered as a continuous variable. Thus continuous variables are those variable that measure the value of an item in real.

A variable is a tool that keep on changing value. This value can vary from one observation to the other. A continuous variable is a variable that take any value possible for observation on the real line. It means all possible positive, negative, fraction etc.

There are uncountable values that exist between two numbers like 1 and 2 for e.g. 1.00, 1.01, 1.001. Following examples will give you more insight about continuous variable. These are:

- Time taken by a person to complete a task.
- Height of a person.
- Wind speed.
- Dust particles in the air.
- Cost of an equipment/object.
- Average speed of bike.
- Mileage of car.

After understanding the definition of the variable, it is easy for one to understand independent and dependent variables as these variables are used in the most of the studies.

Independent Variable

A variable is said to be an independent variable if there is little change in its value. The values of other variable change but when the value of other variables change then there is no change in the value of this variable. Hence, that variable whose value is not affected by any other variable is called an independent variable.

For example: Algal density is a variable and its value determine the quantity of Chlorophyll-a that is used as an indicator of lake water quality. But change in Chlorophyll-a does not have an impact on the value of algal density. In this case, algal density is considered as an independent variable.

Dependent Variable

A variable is said to be dependent variable if its value changes due to change in other variable. The variable that influences the value of this dependent variable is called independent variable.

One can see that there are quantities like algal density, poor water quality and chlorophyll-II that is used as an indicator of lake water quality. Now, chlorophyll-II value are basically dependent on the values of algal density and also on the quality of lake water. So basically, chlorophyll-II is dependent variable and others are independent variables.

Comparison between Different Scales of Data

Nominal Scale

It is a level of measurement that is lowest among all scales and is mostly used with the data that are qualitative in nature rather than quantitative. Some of the important features of nominal scale are mentioned below:

- In nominal scale, the data is divided into different groups or categories.
- These groups or categories are assigned units i.e numbers and objects are determined through the category or group to which they belong.
- In a nominal scale, one can assign different names to the category to which they belong and for analysis purpose numerical values are used to represent these names.
- It is important to note that because the units of a nominal scale are categories, hence no arithmetic operations can be applied between the units of a nominal scale.
- Also there is no quantitative relationship between the categories.
- A basic features of the nominal scale is that of equivalence.
- Every member of a given category or group are same on the basis of classification.
- Experimenter often used nominal scale for categorizing the data into groups and also to count number of observation in a particular group and draw inference based on these values.
- These frequency value also help one in comparing the number of items within each group or category.
- A nominal scale does not possess any mathematical features like equal interval, absolute zero point etc. It allows categorization of objects into mutually exclusive categories.

Ordinal Scale

Ordinal Scale is the next higher scale after nominal scale. It is also used for qualitative data but it has some other features than nominal scale like categorizing data with order or ranking:

- It possesses a low level of mathematical property like one cannot apply arithmetic operation on the data values.
- One can rank the category based on whether they possess more, less or the same amount of object being measured.
- Although with ordinal scale one can compare the objects in terms of "better than", "equal to" or "less than" but it says nothing about the magnitude of difference between adjacent units on th scale.

- It does not have the property of equal interval between consecutive units.
- It does not tell about the absolute level of the units for e.g. all units belong to same higher category or lower category. Thus one cannot give any statement about the data in absolute term.

Interval Scale

It is higher scale than both nominal and ordinal scale. It is used for the qualitative data but one can apply arithmetic operation like multiplication and division on the data having interval scale. Some of the features of interval scale are mentioned below:

- The interval scale possesses the properties of magnitude and equal interval between consecutive units but does not have an absolute zero point. There are equal amount of the variable being measured between adjacent units on the scale.
- The interval scale possesses the properties of the ordinal scale and nominal scale.
- In an interval scale, equal differences between the numbers on the scale represent equal differences in the magnitude of the variable.
- It follows that greater differences between the numbers on the scale represent greater difference in the magnitude of the variable being measured.
- Also smaller difference between the numbers on the scale represent the smaller difference in the magnitude of the variable being measured.

The difference between the interval and ordinal scale is that in ordinal we can order the objects based on the characteristics but difference between the numbers has no quantitative significance but in interval scale one can order the observation and also take interval difference that has quantitative significance.

Ratio Scale

Ratio Scale is the highest level of all measurement scales. It has all the properties all nominal, ordinal and interval scale and other features that differentiate this scale from other scales. These are:

- Ratio scale holds the property of absolute zero.
- It means that a ratio sale data can be compared on the basis of ratio. Without absolute zero property it is not possible to calculate ratio of the data and give interpretation.
- In ratio scale, equal differences between the numbers on the scale represent equal differences in the magnitude of the variable.
- Any object with zero value means complete absence of the characteristics in the object.
- In interval scale, one cannot compute ratio due to unavailability of property of absolute zero. For example, one cannot compute the ratio of different temperature values in a room as it is calculated on an interval scale but in ratio scale one can do this.

Environmental Data

There are many important aspect to the collection of environmental data relating to study design and sampling method that influence the type and sampling method that influence the type and strength of statistical inferences that can made.

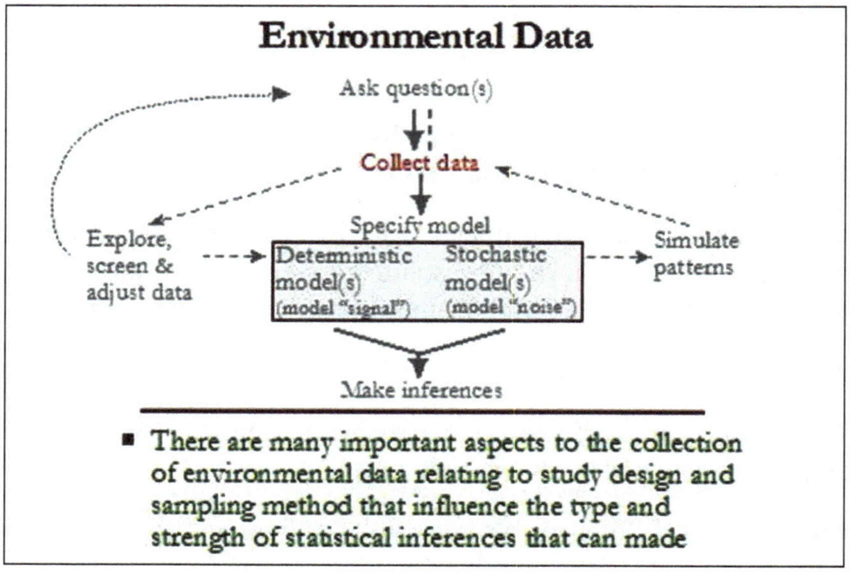

Environmental Data Samples and Population

- We usually wish to make in inferences about a population (statistical, not biology), which is defined as the collection of all the possible observation of interest (size=N).
- We usually collect only a subset of observation from the population called a sample (size=n).
- We infer characteristics of the population from the sample; e.g., estimate parameters, test hypotheses, compare model, and make prediction.

What to Measure

- Types of variables: Independent vs dependent, interdependent.
- "Scale" of data: Continuous, discrete.
- "Type" of data: Count, proportion, binary, time at death, time series, etc.

Types of Variables

Independent versus Dependent

In most cases, we are interested in relating one more independent variables to one or more dependent variables:

- Independent variables: Typically the variable being manipulated or change; controlled or selected by the experimenter to determine its relationship to an observed phenomenon (i.e., the dependent variable); also known as "X","predictor," "regressor," controlled," "maniplated," "explanatory," "exposure,"and/or "input" variable.

- Dependent variable: The observed result of the independent variable being manipulated; usually cannot be directly controlled; also known as "Y", "response" "regressand" "measured," "observed," "responding," "explained," "outcome," "experimental," and/or "output" variable.

Interdependent

In some cases we are interested in a single set of interdependent and dependent:

- Interdependent variables: A set of related variables that are presumed to covary in a meaningful way.

Example:

Table: Sites by species 2-way data matrix.

Sites	Species			
	A	B	C	D
1	1	9	12	1
2	1	8	11	1
3	1	6	10	10
4	10	0	9	10
5	10	2	8	10
6	10	0	7	2

Scale of Variables

- Refers to the scale of measurement or observation (although scale has others dimension) of any variable, dependent or independent.
- Function of the intrinsic nature of the variable and the researcher's choice of how to quantify the variable.
- Affects the form of the statistical model (if dependent variable) and details of the model (if independent).

Scale of Data

Qualitative

- Categorical (nominal, e.g., A, B, C).

Quantitative

- Discrete (ordinal, e.g., ranks, counts).

- Continuous:
    - Interval (arbitrary zero, e.g., temperature).
    - Ratio (true zero, e.g.,mass).

Types of Data

- Refers to the form of the dependent or response variable (not the independent or predictor variable).
- Function of both the scale of the dependent (e.g., continuous versus categorical) and the in which the data was collected owing to the study design.
- Largely determines the overall form the statistical model (model especially regarding the stochastic component).
- Continuous.
- Count:
    - Simple count.
    - Categories.
- Proportion.
- Binary.
- Time to death/failure.
- Time series.
- Circular.

Continuous Data

- Data in which the observation can be measured on a continuum or scale; can have almost any numeric value; can be meaningfully subdivided into finer and finer increments, depending upon the precision of the measurement system.

Examples:

- Temperature.
- Mass.
- Distance.

Some Method:

- Regression.
- Analysis of variance.

Count Data

Data in which the observation can take only the non – negative integer values { 0, 1, 2, ... } and have no upper bound, and where these integers aries from counting rather than ranking.

1. Simple counts:

Plot #infected,

1	2
2	11
3	7
...	...

2. Categorical data:

Species	Town A	B
Sugar maple Red maple Norway maple	4 2 21	9 3 43

Examples:

- # territories.
- # detection in each habitat type.

Some Methods:

- Log–linear models.
- Contingency tables.

Proportion Data

Data in which we know how many of the observation are in one category (i.e., an event occurred) and we also know how many are in each other category (i.e., how many times the event did not occur).

Example:

- Percent mortality.
- Percent infected.
- Sex ratio.

Some Methods:

- Logistic regression.

Trial size	#Infected	#Not infected
10	8	2
15	11	4
12	9	3
...	...	...

Binary Data

Data in which the observations can take only one of two values; useful when you have unique value of one or more explanatory variables for each and every observational unit (which is likely to be the case only with continuous explanatory variables).

Example:

- Present or absent.
- Dead or alive.
- Male or female.

Some Methods:

- Logistic regression.

Individual	Infected
1	0
2	1
3	1
...	...

Time to Death/Failure

Data that take the form of measurement of the time to death, or the time to failure of a component; each individual is followed until it dies (or fails), then the time of death is recorded.

Examples:

- Animal/plant longevity.
- Snag fall.
- Roof failure (leakage).

Some Methods:

- Survival analysis.

Individual	Time to death
1	7
2	10
3	1
....	...

Time Series

Sequence (vector) of data points, measured typically at successive times (or locations), spaced at (often uniform) time (or space) intervals.

Examples:

- Population size.
- Annual temperature.

Some methods:

- Autocorrelation.
- Spectral analysis.
- Wavelet analysis.

Time	Measurement
1	0.07
2	1.20
3	0.61
...	...

Circular Data

Data is which the observation are circular in nature; where the beginning and end of the sequence is the same.

Examples:

- Aspect.
- Day of year.

Some Methods:

- Circular stats.

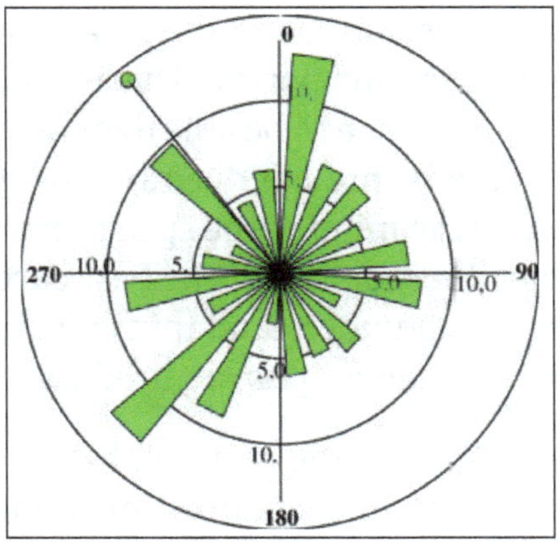

Sampling Design

Next, we need to determine where, when, and how often to collect the data:

- Scale: Matching observational/experimental units to the ecological question.
- Randomization: Obtaining an unbiased sample.
- Replication: Minimizing uncertainty.
- Control: Accounting for important sources of variation.

Representation of Data

Statistics is a science that is based on description of data either numerically or diagrammatic or other way. This science is used to extract information from the data based on the objective. Data can be in quantitative form as well in qualitative form. Data are collected from the resources of study and are available in raw form. After collecting and editing the data the next stage is to organize the data. Classification and tabulation of the data are among the most important tools for the precise, clear and comprehensible representation of the data. However, sometimes these forms of presentation are not appealing to the common person. Due to technicality involved in these forms, it may not be interesting for a common person to be able to understand the things in a simple manner. Another way to represent the data is through diagram and graphs that present the data into attractive manner that appeal more to the mind of the spectators. These forms are more attractive, fascinating and impressive than the other methods. The best part of diagrammatic representation method is that even a layman person can understand this without any previous knowledge of statistics. This is the reason that diagrams and graphs are used to give basic education to the kids.

Another important feature of the diagrammatic and graphical representation is that it saves lot of time as these are easy to build up and one can draw meaningful inference from them. These methods are able to present that information that might be lost amid the details of classification and tabulation of data. These methods also facilitate the person for comparing the value of two or more sets of data. Graphs and charts are used to clarify a complex problem and reveal the hidden facts that are not clear from the tabular form. Hence, graphs and diagrams are important not only for the representation of the data but for visual comparison of two or more datasets.

Frequency Distribution

In English language, frequency means the rate of occurrence of something in a repetitive manner in a particular time interval or time frame. Statistics is concerned with the extraction of information from the numbers collected in a raw form from the study.

In practical life, variables are used to represent these numbers in case of quantitative data like in a study of sales of luxury cars in a particular region in a year. One can consider "the sale of a car in a day" as a variable and note down the sale of a car in a day by x_i variable where I = 1, 2,.....365. Here the values, $x_1, x_2, \ldots .x_{365}$ represent the sales of a car in the first day of a year or financial year and similarly second day of the year and so on. The values of $x_1, x_2, \ldots, x_{365}$ are known as the values of the variable.

Frequency of a value of the variable means the number of times a same value is repeated in the whole dataset. For example, if we assume that the no. of sales of a car are 2 units on 10th day, 3 units on 20th day, 2 units on the 45th day, 4 units on the 70th day of the year and so on. Here, 2 units appear two times, 3 and 4 units appear single time from the available information. So, from the available information the frequency of sale of 2 cars per day is two and other has only one. With an increase in the available information, one can construct frequency table that represent the repetition of the values in a dataset.

For example, let us consider the observation of sales of car in the month of February (28 values). These are 2, 3, 1, 2, 4, 5, 3, 2, 1, 3, 5, 6, 1, 2, 4, 2, 3, 5, 1, 3, 4, 2, 3, 5, 4, 1, 4, 2.

Observations	Frequency
1	5
2	7
3	6
4	5
5	4
6	1
Total	28

In the above table, we observe that for seven days two units of cars are sold, for six days three cars are sold and so on. Hence one can form the frequency table by counting the same observations in the data set this means frequency of a value of a variable is the number of times it occurs in a given series of observations. The table that represents the frequency of the value side by side is called frequency table. There are two forms of frequency distributions- Ungrouped frequency distribution and Grouped frequency distribution.

To understand the difference between two methods, we consider a data set of IQ score of 30 students and construct ungrouped and grouped frequency distribution.

IQ Scores of students				
35	65	58	79	41
45	47	87	85	47
51	81	88	76	83
38	62	94	79	84
61	74	86	64	85
54	78	73	71	77

Ungrouped Frequency Distribution

Ungrouped frequency distribution table shows the frequency of the values in the dataset on individual basis. The table below is an example of ungrouped frequency distribution as the first column in the table represents the observations and second column shows corresponding frequency.

Ungrouped Frequency Table			
IQ score	Frequency	IQ Score	Frequency
35	1	74	1
38	1	76	1
41	1	77	1
45	1	78	1
47	2	79	2
51	1	81	1
54	1	83	1
58	1	84	1
61	1	85	2
62	1	86	1
64	1	87	1
65	1	88	1
71	1	94	1
73	1		

From table, one can observe that 2nd column and 4th column show the frequency values of the dataset. Hence in ungrouped frequency distribution, the values of the dataset are shown on individual basis.

Group Frequency Distribution

In this method, the values of the variables are shown in the group or interval. In the following table, observations from table are used to present them in group frequency distribution. The smallest observation in the dataset is 35 and maximum value is 94. In table, we consider width of the interval as 10 and lowest class value as 30 and so on till we cover the maximum value in the dataset.

From table, we can see 70-80, 80-90 intervals have maximum frequency value that is 8. Hence we can conclude that IQ score of the most of the students lie between 70 to 90.

On the comparison of table, one can see that group frequency distribution visualize the important characteristics of the data in a simple and understandable manner about the tendency of IQ score of students over ungrouped frequency distribution.

Grouped Frequency Distribution	
IQ Score	Frequency
30-40	2
40-50	4
50-60	3
60-70	4
70-80	8
80-90	8
90-100	1

Steps for Forming a Group Frequency Distribution

Many techniques are used for the formation of the group frequency distribution. For group of observations, we divide the data into class intervals and difference between upper and lower interval is called the width of the class interval. There are few points that must be kept in mind while preparing a group frequency table:

- Class type: One should define class type in a very clear manner. It should be exhaustive and mutualy exclusive so that variable's value must be assigned to only one class in the table.
- Class Intervals: It means how many intervals should be formed for the available data. Number of intervals depend on the things like number of observations in the data, its magnitude value, level of precision and further analysis of the data. The most common formula that is used for the determination of the interval in the group frequency distribution is Sturge's rule:

  $$k = 1 + 3.322 \log_{10} N$$

  where k is the number of classes and N is the total number of observations in the data. This rule is used for correct determination of intervals in the data and it is further used for the determination of the width of the class interval.

- Width of class interval: Width of the interval means the difference between the lower limit and upper limit of the interval. The width of the interval is defined through the formula that is:

  $$h = \frac{\text{Range}}{\text{Number of classes}}$$

  where h denote the width of the class interval and range is defined as the difference between the highest and lowest value of the data set.

- Class limits methods: There are different methods that are used for the classification of the data set on the basis of class interval. The limit consists of two numbers that are used for the purpose of tallying observations into various classes. There are two different methods for the classification of the data on the basis of class intervals. These are:- (a) Inclusive method and (b) Exclusive Method:
    - Inclusive method: In inclusive method, the upper limit of a class interval is considered in the interval itself and is not related with the next class. For example, in inclusive method, the class limits are 0-4, 5-9, 10-14, 15-19 and so on. Hence one can see that both the upper and lower limits are included in the class and thus it is termed as inclusive. The main drawback of this method for continuous data observations for example if data value is 4.5 then with this method one cannot tabulate or assign the value to any interval.
    - Exclusive method: In this method, the data are classified into class interval of such time that upper limit of one interval is the lower limit of next succeeding class interval. For example, in exclusive method the class limits should be of such type that is 0-5, 5-10,10-15 and so on. Hence all those values that are less than 5 are considered in first interval and all those data values that are above than 5 but less than 10 are counted in second interval. Hence in exclusive method, the problem of inclusive method is taken care of.
- Mid value or Mid points: Mid value is calculated by taking the sum of upper and lower limit of the interval and dividing that sum by 2. This value is used as a representative value of the class interval and it is used for evaluation of mean, median, mode and higher moments of the data.

Cumulative Frequency Distribution

Frequency distribution counts the occurrence of the same value in the data but sometime one is interested in the number of observations that are small or greater than a given value. In such type of situation, one has to calculate the accumulated frequency less than or greater than some specified value. This accumulated value is known as Cumulative frequency distribution.

The frequency of observations till a given value is considered as less than cumulative frequency and the frequency of observations that are greater than a value is called more than cumulative frequency.

Using the same observation as given in table. The cumulative frequency distribution for both more and less than are given in the following table:

IQ Score	Frequency	Less than Cumulative Frequency	More than Cumulative Frequency
30-40	2	2	30
40-50	4	6	28
50-60	3	9	24
60-70	4	13	21
70-80	8	21	17
80-90	8	29	9
90-100	1	30	1

From the above table, one can see the less than and more than cumulative frequency values of the data. These values are further used for graphical representation of the data. For example, ogive curve of more than and less than type use these values for plotting on the axis.

Use of Diagrammatic and Graphical Presentation of the Data

Diagrammatic and Graphical presentation of the data are useful in practice due to the following reasons. These are:

- The information that we acquire from the graphical and diagrammatic representation of the data is easy to understand even for a layman person due to its simplicity.
- People are more interested in graphical presentation of facts then just numbers due to eye caching effect of diagram or pictures.
- Graphs and picture can simplify the complexity of the data that cannot be easily be understood with the figures.
- With the graphical presentation, one can easily compare the statistical data relating to different time and places to bring out the hidden facts and relationship among the statistical variables.

There are some limitations of diagrammatic and graphical representations like they do not show the details behind the numbers that can only be shown from the table in a better way. A single diagram or graph does not have a great importance rather than it is used for comparison purpose with other diagram or graph.

Difference between Diagrams and Graphs

There are few rules based on them, one can differentiate between graphs and diagram but these rules are not standard for all so there is scope of changes in these rules among different persons. These rules are:

- Diagram are plotted on the paper while graphs are plotted on a paper called graph paper graphs helps in the study of mathematical or numerical relationship between variables but in diagram precise relationship among variables are not discussed.
- In diagram, different tools like bars, rectangles, circles etc are used to present the information in the data. Whereas in graphs, different tools like lines, dots etc are used to present the data.
- Diagrams only give approximate information regarding the data as this information will not be used further for analysis purpose. On the other hand, graphs give more precise, accurate information about the data and they are used for further analysis purpose.
- Diagrams are used for the presentation of the categorical and geographical information in the data. On the other hand, graphs are used for the presentation of the time series and frequency distribution.

- Diagrams are more eye catching than graphs. Also diagrams are used for the understanding of the layman person but graphs are used by experts from the field for the further analysis of the information.
- Graphs are easier to build than comparative to the diagram.

There are few points based on which one keep in mind while constructing the diagrams. These are:

- Diagram gives only a pictorial representation of the quantitative data for rough guesses.
- It can only be used for homogenous data.
- It is not reliable to make further inference about the data.

So, basically diagram are used for the graphical interpretation purpose only. One cannot use it to find out reasons or inference from the data.

While constructing diagrams, there are some general rules that should be followed. These are:

- An appropriate diagram can only present the data in a better way. Thus, it is essential to choose the right diagram for the data that need expertise as well as knowledge. It may be possible that due to inappropriate selection of diagram the interpretation might be wrong that can lead to unbearable results.
- It is also important that a diagram should have an appropriate title corresponding to the nature of the data. With an appropriate title, a person can understand the main idea in the diagram.
- It should be constructed in such a way that it portray all the relevant information within an allotted space. So, it should be appropriate in terms of size and consistent in terms of dimensions.
- It should be neat, clean with footnotes and proper indexing that will attract the interest of the common man.

Graphical Representation of Data

There are certain guidelines for an attractive and effective graphic presentation of data and information. These are as follows:

- Suitable Title – Ensure that you give a suitable title to the graph which clearly indicates the subject for which you are presenting it.
- Unit of Measurement – Clearly state the unit of measurement below the title.
- Suitable Scale – Choose a suitable scale so that you can represent the entire data in an accurate manner.
- Index – Include a brief index which explains the different colors and shades, lines and designs that you have used in the graph. Also, include a scale of interpretation for better understanding.

- Data Sources – Wherever possible, include the sources of information at the bottom of the graph.
- Keep it Simple – You should construct a graph which even a layman (without any exposure in the areas of statistics or mathematics) can understand.
- Neat – A graph is a visual aid for the presentation of data and information. Therefore, you must keep it neat and attractive. Choose the right size, right lettering, and appropriate lines, colors, dashes, etc.

Merits of a Graph

- The graph presents data in a manner which is easier to understand.
- It allows us to present statistical data in an attractive manner as compared to tables. Users can understand the main features, trends, and fluctuations of the data at a glance.
- A graph saves time.
- It allows the viewer to compare data relating to two different time-periods or regions.
- The viewer does not require prior knowledge of mathematics or statistics to understand a graph.
- We can use a graph to locate the mode, median, and mean values of the data.
- It is useful in forecasting, interpolation, and extrapolation of data.

Limitations of a Graph

- A graph lacks complete accuracy of facts.
- It depicts only a few selected characteristics of the data.
- We cannot use a graph in support of a statement.
- A graph is not a substitute for tables.
- Usually, laymen find it difficult to understand and interpret a graph.
- Typically, a graph shows the unreasonable tendency of the data and the actual values are not clear.

Types of Graphs

Graphs are of two types:

- Time Series graphs.
- Frequency Distribution graphs.

Time Series Graphs

A time series graph or a "histogram" is a graph which depicts the value of a variable over a different point of time. In a time series graph, time is the most important factor and the variable is related to time. It helps in the understanding and analysis of the changes in the variable at a different point of time. Many statisticians and businessmen use these graphs because they are easy to understand and also because they offer complex information in a simple manner.

Further, constructing a time series graph does not require a user with technical skills. Here are some major steps in the construction of a time series graph:

- Represent time on the X-axis and the value of the variable on the Y-axis.
- Start the Y-value with zero and devise a suitable scale which helps you present the whole data in the given space.
- Plot the values of the variable and join different point with a straight line.
- You can plot multiple variables through different lines.

Line Graph

You can use a line graph to summarize how two pieces of information are related and how they vary with each other.

Advantages:

- You can compare multiple continuous data-sets easily.
- You can infer the interim data from the graph line.

Disadvantages:

- It is only used with continuous data.

Use of a False Base Line

Usually, in a graph, the vertical line starts from the Origin. However, in some cases, a false Base Line is used for a better representation of the data. There are two scenarios where you should use a false.

Base Line:

- To magnify the minor fluctuation in the time series data.
- To economize the space.

Net Balance Graph

If you have to show the net balance of income and expenditure or revenue and costs or imports and exports, etc., then you must use a net balance graph. You can use different colors or shades for positive and negative differences.

Frequency Distribution Graphs

Let's look at the different types of frequency distribution graphs:

Histogram

A histogram is a graph of a grouped frequency distribution. In a histogram, we plot the class intervals on the X-axis and their respective frequencies on the Y-axis. Further, we create a rectangle on each class interval with its height proportional to the frequency density of the class.

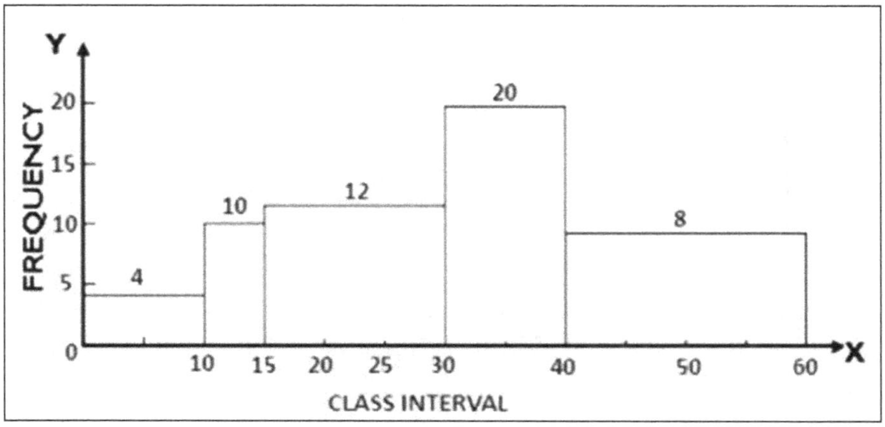

Frequency Polygon or Histograph

A frequency polygon or a Histograph is another way of representing a frequency distribution on a graph. You draw a frequency polygon by joining the midpoints of the upper widths of the adjacent rectangles of the histogram with straight lines.

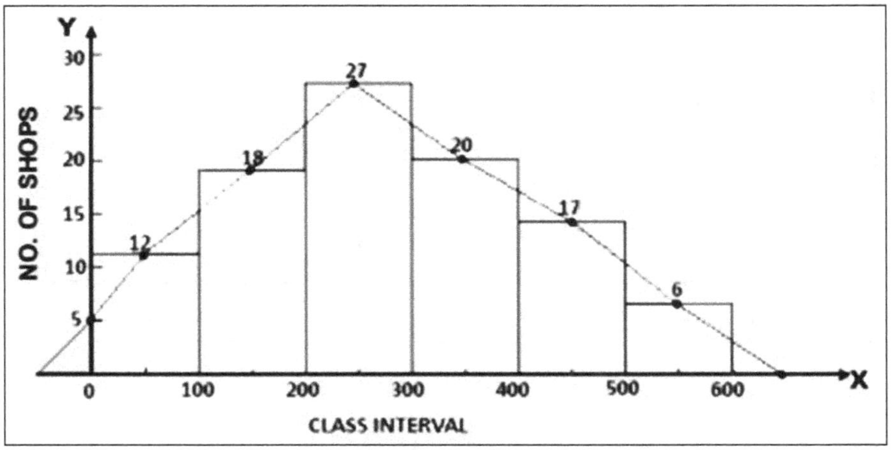

Frequency Curve

When you join the verticals of a polygon using a smooth curve, then the resulting figure is a Frequency Curve. As the number of observations increase, we need to accommodate more classes. Therefore, the width of each class reduces. In such a scenario, the variable tends to become continuous and the frequency polygon starts taking the shape of a frequency curve.

Cumulative Frequency Curve or Ogive

A cumulative frequency curve or Ogive is the graphical representation of a cumulative frequency distribution. Since a cumulative frequency is either of a 'less than' or a 'more than' type, Ogives are of two types too – 'less than ogive' and 'more than ogive'.

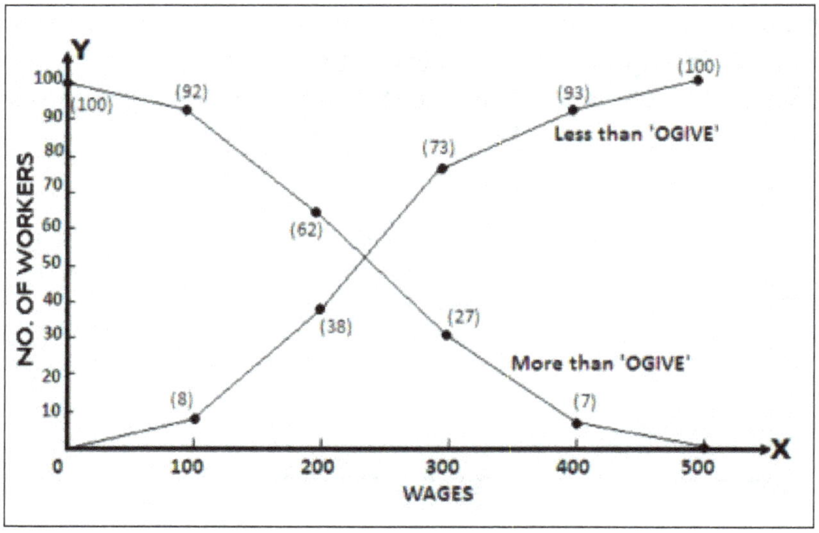

Scatter Diagram

A scatter diagram or a dot chart enables us to find the nature of the relationship between the variables. If the plotted points are scattered a lot, then the relationship between the two variables is lesser.

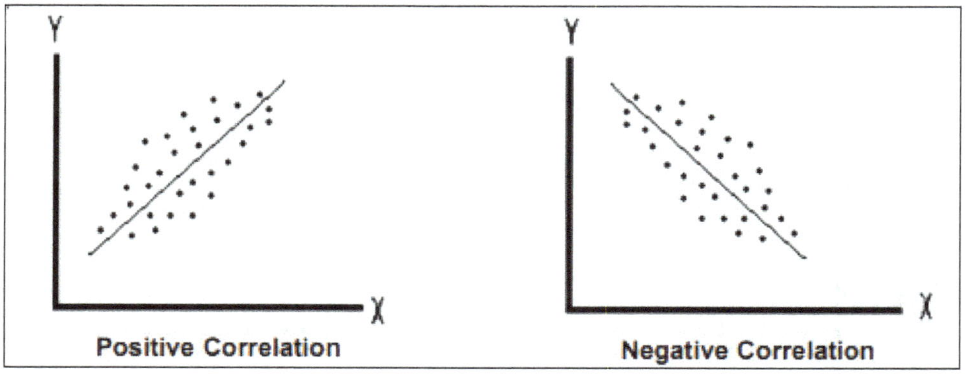

Central Tendency Measures

In Statistics, three measures are considered generally that is mean, median and mode. Mean measure is further divided into three categories (i) Arithmetic Mean (ii) Geometric Mean and (iii) Harmonic Mean.

The relationship between different measures are shown below:

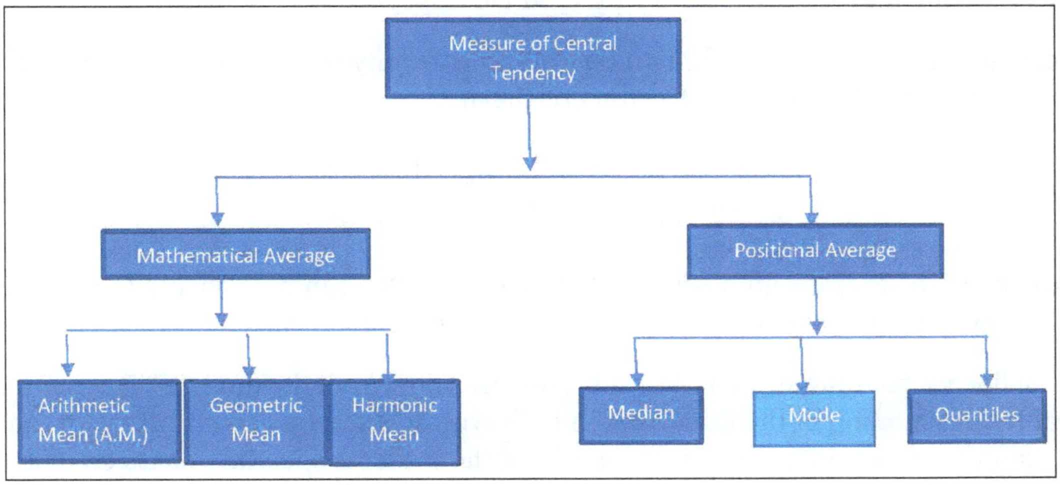

From the above measures, the arithmetic mean, median and mode are widely used. On the other hand, geometric and harmonic mean are used in particular situations.

Mathematical Averages

Arithmetic Mean (A.M.)

Arithmetic Mean is among the most widely used measure of central tendency for a single representation of observations. The formula for the evaluation of the arithmetic mean (A.M) is given as:

$$\text{Arithmetic Mean } (\text{A. M.}) = \frac{\text{sum of all obervations}}{\text{number of observations}}$$

Let $x_1, x_2, \ldots\ldots, x_n$ be n observation is a data set. The Arithmetic mean of the data is denoted as $\bar{x}$.

$$\bar{x} = \frac{\sum_{i=1}^{n} x_i}{n}$$

For example, let 23, 24, 25, 35, 32 be 5 class size of different sections of 9th class in a private school. We want to know about the average class size in 9th class. To calculate this we first calculate the sum of all class sizes and divide it by no of sections. The arithmetic mean is 27.8 ≈ 28. It means that there is approximately 28 students on average basis in each class.

This method cannot be applied if the no. of the elements in the data set are very large. For the convenience purpose frequency tabulation method is used.

Let $x_1, x_2, \ldots, x_n$ have frequencies $f_1, f_2, \ldots, f_n$ respectively that means that in the complete data set x_1 appears f_1 times and x_2 appears f_2 times and so on.

Hence the formula for A.M. in this case will be,

$$\bar{x} = \frac{\{f_1x_1 + f_2x_2 + \cdots + f_nx_n\}}{f_1 + f_2 + \cdots + f_n}$$

This method of calculating Arithmetic Mean in term of frequencies is known as Weighted Arithmetic Mean. As we are multiplying the x_i's with their corresponding frequencies (f_i).

An example consists of no. of child in 50 families in a locality is presented below will give you a better understanding of the weighted arithmetic mean.

No. of child	0	1	2	3 or above
Families	10	15	20	5

Now to calculate the average number of child in a family. One can use weighted arithmetic mean and the value is $1.4 \approx 1$. Hence on average there is a single child in each family.

Also in real life, we have large amount of data so as we know from the module "Diagrammatic and Graphical Representation of the Data" that in such type of cases, one can construct frequency table. For example, we are interested in the marks of the students spending hours on daily basis to learn through online program. A artificial data was available of 1000 students and information is given in the following table as:

Time Spent (in hours)	No. of students
0-4	690
4-8	285
8-12	20
12-16	5

Now if one is interested to find the average no of hours spend on average basis using A.M. then the formula is:

$$\overline{x} = \frac{\{f_1 x_1 + f_2 x_2 + \cdots + f_n x_n\}}{f_1 + f_2 + \cdots + f_n}$$

but here $x_1, x_2, ..., x_n$ are the midpoint of class interval. So the arithmetic mean of the data can be found as,

$$= \frac{2*690 + 6*285 + 10*20 + 14*5}{690 + 285 + 20 + 5} = \frac{3366}{1000} = 3.360$$

Hence from the above example, we can conclude that 3.36 hours or 3 hours and 6 minutes on average are spent by the students on daily basis.

Merits

- A.M. can be easily understood by a person and does not require any particular need to explain it.
- The calculation of the A.M. is very simple and does not require any calculations.
- It is derived without ordering the data like other measure of central tendency.

- It is considered as comparative very stable and good method of central tendency in comparison to other measures.
- If one has large amount of data then it is considered as a good method of central tendency as the unusual observations in one direction can be offset by the unusual observations in other direction.

Demerits

There are also certain shortcoming of any measure and A. M. is not an exception. There are some important demerits of A.M given below:

- A.M. cannot be obtained through guesses like other measure mode etc. whether it is not considered as a major demerit but still it is. Also median can be located graphically but A.M. cannot.
- A.M. cannot be found for the data when one has missing observation in the data. Also for a small set of data it is highly sensitive the extreme values. In such cases, A.M. will lead to a wrong conclusions.
- A.M. cannot be evaluated for the qualitative characteristics that cannot be measured, like sex, level of illness etc. In such cases it is suitable to use other measure of central tendency like median.
- It may be possible that a value calculated through A.M. is difficult to interpret. For example, one can recall the value of A.M. evaluated from table which is 1.4 that has no interpretation as child is considered in unit not in fraction. So the 1.4 child is meaningless.
- It is possible that it may lead to false conclusions.

At the last one can say that A.M. is considered as a simplest measure of central tendency but it will mislead you when data are of heterogeneous form i.e. non-homogeneous.

Geometric Mean

Geometric Mean is another method to find out the mean of n observations. It is evaluated through the following method:

Let us suppose $x_1, x_2, \ldots, x_n$ are n observations, then,

$$G.M. = \left(x_1, x_2, \ldots, x_n\right)^{\frac{1}{n}}$$

If $x_1, x_2, \ldots, x_n$ have frequencies $f_1, f_2, \ldots, f_n$ respectively, for evaluating the G.M. the method is,

$$G.M. = \left(x_1^{f_1}, x_2^{f_2}, x_3^{f_3}, \ldots, x_n^{f_n}\right)^{\frac{1}{N}}$$

where, $N = \Sigma f_i$ is the sum of all observation's frequency and is called the total frequency.

Hence, one can use the above method to find the G.M. of the observations. It is difficult to find out the G.M. by using this formula or one can use the above formula when the no of observations is very small. In this case, one can use log on both side of the formula. Then the modified form the above formula are given below respectively,

$$\log G.M. = \frac{1}{n}\sum \log(x_i)$$

$$\log G.M. = \frac{1}{N}\sum f_i \log(x_i)$$

Now, if one is using the formula of equation above, then one has to take antilog at the end to find out the G.M. value.

$$G.M. = \text{Anti}\log\left(\frac{1}{n}\sum \log(x_i)\right)$$

$$G.M. = \text{Anti}\log\left(\frac{1}{N}\sum f_i \log(x_i)\right)$$

Steps for the calculation of the G.M. for simple data using equation above:

- Calculate log (base 10) of the observations.
- Take the sum of all log of observations i.e. $\sum \log(x_i)$.
- To find the log G.M., divide $\sum \log(x_i)$ by total no. of observations.
- Take antilog (base 10) of $\frac{1}{n}\sum \log(x_i)$ and this is G.M. of the required data.

Example: Find out the G.M. of the following data given as 34, 57, 37, 43, 52, 67, 54, 56.

In table, the log values are shown in second column.

Observations	Log (x_i)
34	1.5314789
57	1.755874
37	1.5682017
43	1.6334685
52	1.7160033
67	1.8260748
54	1.7323938
56	1.748188
No. of obs =8	13.511684
log G.M.	1.6889605
G.M.	48.86079

From the above data, the required value is calculated as 48.86079.

Now to find the G.M. of the data when the data set has frequencies with them. There are further two cases:

- Discrete case (Ungrouped Frequency Series).
- Continuous case or Class Interval (Grouped Frequency Series).

In the first case, observations are given with their corresponding frequencies. In the second case, as data is given in the class interval. In these two cases the G.M. is calculated in a different manner.

- Discrete Case: G.M. for the discrete case is considered first. One has to follow some steps given below to calculate the G.M.

Steps are:

- Calculate log (base 10) of the observations.
- Multiply the log values with the corresponding frequencies values and take their sum i.e, $\sum f_i \log(x_i)$.
- Divide the term i.e $\sum \log(x_i)$ by N that is the total frequency.
- Take anitlog (base 10) of $\frac{1}{N}\sum f_i \log(x_i)$. This will give you the value of G.M for frequency data.

Example: Find out the G.M. of the following data given as,

Observations	Frequency
38	7
43	9
46	10
49	6
51	4
54	8
67	3
78	5

Solution:

Observations	Frequency	log (x_i)	freq*log(x_i)
38	7	1.5797836	11.05848518
43	9	1.6334685	14.7012161
46	10	1.6627578	16.62757832
49	6	1.6901961	10.14117648
51	4	1.7075702	6.830280704
54	8	1.7323938	13.85915008
67	3	1.8260748	5.478224408
78	5	1.8920946	9.460473013
	N = 52		88.15658428
	log (G.M.)		1.695318928
	G.M.		49.58141632

In the table, data values are given in the first column and their corresponding frequencies in the second column. Logarithm of observations are calculated in the third column. Fourth column represents the values of product of logarithm values corresponding to their frequencies. 88.15658428 represents $\sum f_i \log(x_i)$ and 1.695318928 is calculated by dividing 88.15658428 by N i.e. 52 and 49.5841632 is the anitlog of 1.695318928. Hence the G.M. is calculated as 49.5841632.

- Continuous case or Class Interval: If the data is of continuous form then one has to first construct frequency table using class intervals. Then only one can find out the G.M. for the continuous data. There are few steps for calculate G.M. in these cases. These steps are:
    - First calculate the mid points (m_i) of the class intervals.
    - Calculate log (base 10) of the mid values of each class interval i.e. log (m_i).
    - Multiply the log values with the corresponding frequencies values and take their sum i.e, $\sum f_i \log(m_i)$.
    - Divide the term i.e $\sum f_i \log(m_i)$ by N that is the total frequency.
    - Take anitlog (base 10) of $\frac{1}{N}\sum f_i \log(m_i)$. This will give you the value of G.M for frequency data.

Example: Find the G.M. of the following data.

Class Interval	Frequency
0.0-5.0	4
5.0-10.0	6
10.0-15.0	8
15.0-20.0	5
20.0-25.0	7
25.0-30.0	3

Solution:

Class Interval	mid value (m_i)	Frequency	log (m_i)	log (m_i) *freq
0.0-5.0	2.5	4	0.397940009	1.591760035
5.0-10.0	7.5	6	0.875061263	5.25036758
10.0-15.0	12.5	8	1.096910013	8.775280104
15.0-20.0	17.5	5	1.243038049	6.215190243
20.0-25.0	22.5	7	1.352182518	9.465277627
25.0-30.0	27.5	3	1.439332694	4.317998081
		N = 33		35.61587367
			Log G.M.	1.07926889
			G.M.	12.00242219

In the table, first column represents the class intervals and their corresponding frequencies are given in the third column. Mid values are calculated in the second column and logarithm of mid values are calculated in the fourth column. Fifth column represents the values of product of logarithm values corresponding to their frequencies. 35.61587367 represents $\sum f_i \log(m_i)$ and 1.079268899 is calculated by dividing 35.61587367 by N i.e. 33 and 12.00242219 is the anitlog of 1.079268899. Hence the G.M. is calculated as 12.00242219.

Hence, one can use the above steps for calculating the G.M. in continuous case.

As we are using the G.M. as a measure of central tendency one must know where to use this measure to get more appropriate information. Hence it is essential to discuss about the merits and demerits of this measure.

Merits of G.M.

- It is defined in a rigid manner.
- It is based on all the values of the data and cannot be calculated for data with missing values.
- G.M. can be calculated in those cases where only the total product of observations and number of observations are known without having knowledge of individual values.
- G.M. can be used for further mathematical treatment like rations and percentages. Therefore it can be used for further analysis like index number etc.
- The main advantage of G.M. is that it is not affected much by extreme large and small values and it gives more weight to small values in comparative to A.M. which gives more weight to large values.

Demerits of G.M.

There are few demerits of G.M. These are:

- Due to the complexity involved in the calculation of G.M. . It is difficult for a non-mathematical person to understand it.
- The calculation of G.M. requires the knowledge of logarithm that is difficult for a nonmathematical background person.
- G.M. cannot be calculated for the data which contain a value zero.
- G.M. is not useful for the data that have different signs that is some of the observations are positive and other are negative. In such case, G.M. value has no meaning.
- G.M. value may not be the actual value of the observations like other measure of central tendency.
- G.M. is not advisable to use in those cases where small observations must be given more weights and higher observation must be given less weights.

Another measure of central tendency is Harmonic Mean (H.M.). This measure of central tendency has a great importance in science and mathematical field.

Harmonic Mean

Harmonic mean is calculated by first taking the reciprocal of the observations and then taking the reciprocal of arithmetic of these reciprocal observations.

Let us suppose $x_1, x_2, \ldots, x_n$ are n observations in the dataset. Then,

$$\text{H.M.} = \frac{n}{\frac{1}{x_1} + \frac{1}{x_2} \cdots + \frac{1}{x_n}} = \frac{n}{\sum \frac{1}{x_i}}.$$

When observations are given with frequencies i.e. $x_1, x_2, \ldots, x_n$ with $f_1, f_2, \ldots, f_n$ are the corresponding frequency values then the H.M. is evaluated by using the following formula:

$$\text{H.M.} = \frac{f_1 + f_2 + \cdots + f_n}{\frac{f_1}{x_1} + \frac{f_2}{x_2} \cdots + \frac{f_n}{x_n}} = \frac{N}{\sum f_i / x_i}$$

where, $N = \sum f_i$ is the sum of all observation's frequency and is called the total frequency.

Harmonic mean is basically used in the case where rates and ratio are involved like per hour, per month, per litre, per min etc. It is used for to find out the average of different types of rates. These rates are basically used to explain the relationship between two units of opposite nature like as you increase the speed of a vehicle then this shortens the travelling time. Hence H.M. is used to find out the average speed in kilometer per hour for a given distance.

In the following example, H.M. is evaluated to find out the average speed of 6 cars given in kilometer per hour.

Example: Calculate the Harmonic Mean (H. M.) for the series 56, 45, 67, 58, 53, 42.

Solution:

KMPH	Reciprocal of observations
56	0.017857143
45	0.022222222
67	0.014925373
58	0.017241379
53	0.018867925
42	0.023809524
N = 6	0.114923566
H.M.	52.20861322

In table, the reciprocal of observation are given in the second column. 0.114923566 represents the total of reciprocal of the observations and 52.2086 is the H.M. of the data.

Similarly, one can keep the following steps in mind to find out the H.M. for observations given with frequencies.

For ungrouped frequency data. Steps are:

- Find out the reciprocal of the observations.
- Multiply the reciprocal observations with the corresponding frequencies values and take their sum i.e $\sum f_i / x_i$.
- Divide N by the term i.e $\sum f_i / x_i$ where N is the total frequency.
- This will give you the value of H.M. for discrete frequency data.

For grouped frequency data. Steps are:

- First calculate the mid points (m_i) of the class intervals.
- Multiply the reciprocal observations with the corresponding frequencies values and take their sum i.e $\sum f_i / x_i$.
- Divide N by the term i.e $\sum f_i / x_i$ where N is the total frequency.
- This will give you the value of H.M. for grouped frequency data.

Hence, one can evaluate the H.M. for three different situations i.e. simple, discrete frequency and grouped frequency.

Merits:

- H.M. is defined in a rigid manner.
- H.M. is evaluated on all the observations in the data.
- H.M. gives less weight to large values and more weight to small values so this measure is appropriate in those case where such condition is required.
- H.M. is used to measure relative changes and is widely used for the averaging of ratio and rates.

Demerits:

- H.M. is very difficult measure to be understood by a non-mathematical background person.
- H.M. is comparatively difficult to calculate than other measure.
- H.M. cannot be evaluated if the value of the observation is zero.
- H.M. cannot be evaluated when some of the observations are positive and some are negative in the data set.
- H.M. value does not exist in the data set.

Relation between AM, GM and HM

Before we relate the three means in Statistics, which are Arithmetic Mean, Geometric Mean and Harmonic Mean, let us understand them better.

Arithmetic Mean

Arithmetic mean represents a number that is achieved by dividing the sum of the values of a set by the number of values in the set. If $a_1, a_2, a_3,\ldots,a_n$, is a number of group of values or the Arithmetic Progression, then;

$$AM = (a_1+a_2+a_3+\ldots,+a_n)/n$$

Geometric Mean

The Geometric Mean for a given number of values containing n observations is the nth root of the product of the values.

$$GM = n\surd(a_1a_2a_3\ldots a_n)$$

Or,

$$GM = (a_1a_2a_3\ldots a_n)^{1/n}$$

Harmonic Mean

HM is defined as the reciprocal of the arithmetic mean of the given data values. It is represented as:

$$HM = n/[(1/a_1) + (1/a_2) + (1/a_3) + \ldots + (1/a_n)]$$

What is the Relation between AM, GM and HM?

The relationship between AM, GM and HM is given by:

$$AM \times HM = GM^2$$

First, consider a, AM, b is an Arithmetic Progression.

Now the common difference of Arithmetic Progression will be;

$$AM - a = b - AM$$

$$a + b = 2\,AM$$

Secondly, let a, GM, b is a Geometric Progression. Then, the common ratio of this GP is;

$$GM/a = b/GM$$

$$ab = GM^2$$

Third, is the case of harmonic progression, a, HM, b, where the reciprocals of each term will form an arithmetic progression, such as:

$1/a$, $1/HM$, $1/b$ is an AP.

Now the common difference of the above AP is;

$$1/HM - 1/a = 1/b - 1/HM$$

$$2/HM = 1/b + 1/a$$

$$2/HM = (a + b)/ab$$

Substituting equation $a + b = 2\ AM$ and equation $ab = GM^2$ in equation $2/HM = (a + b)/ab$ we get;

$$2/HM = 2AM/GM^2$$

$$GM^2 = AM \times HM$$

Hence, this is the relation between Arithmetic, Geometric and Harmonic mean.

Example:

1. If five times the geometric mean of two numbers 'a' and 'b' is equal to the arithmetic mean of those two number such that a > b > 0, then compute the value of a + ba – b.

Solution:

Arithmetic mean of two numbers is calculated as:

$$AM = \frac{a+b}{2}$$

Geometric mean of the two numbers a and b is,

$$GM = \sqrt{ab}$$

It is given in the question that Arithmetic mean = 5 times the geometric mean,

$$AM = 5\ GM$$

$$\frac{a+b}{2} = 5\sqrt{ab}$$

$$a+b = 10\sqrt{ab}$$

$$(a+b)^2 = 100\ ab$$

$$(a-b)^2 = (a+b)^2 - 4\ ab$$

$$(a-b)^2 = 100\ ab - 4\ ab$$

$$(a-b)^2 = 96\ ab$$

$$a-b = \sqrt{96ab}$$

$$\frac{a+b}{a-b} = \sqrt{\frac{100\ ab}{96\ ab}} = \sqrt{\frac{25}{24}} = 1.021$$

2. Find the harmonic mean of two numbers a and b, if their arithmetic mean is 16 and geometric

mean is 8 provided that a > b > 0. (Hint: Use relation between AM GM HM formula). Also find the values of a and b.

Solution:

Given: AM = 16 and GM = 8

The relation between AM GM HM is given as:

$$AM \times HM = GM^2$$
$$16 \times HM = 8^2$$
$$16 \times HM = 64$$
$$HM = \frac{64}{16} = 4$$

The find the numbers:

Arithmetic mean is given as:

$$AM = a + b / 2$$
$$16 = a + b / 2$$
$$a + b = 32$$
$$a = 32 - b$$

Geometric mean is given as:

$$GM = \sqrt{ab}$$
$$8 = \sqrt{ab}$$
$$64 = ab$$
$$64 = (32 - b)b$$
$$64 = 32b - b^2$$
$$b^2 - 32b + 64 = 0$$

Calculation of Mean for Different Series

Calculation of Mean for different series using different methods.

Individual Series

Direct Method

$$\bar{X} = \frac{\Sigma X}{N}$$

where, ΣX = Sum of all observations and N = Total number of observations.

For example:

The following data shows the weekly income of 10 families. Calculate the arithmetic mean by direct method and interpret the result:

Family	A	B	C	D	E	F	G	H	I	J
Weekly Income (in ₹)	850	700	100	750	5,000	80	420	2,500	400	360

Computation of Arithmetic Mean by Direct Method:

Family	A	B	C	D	E	F	G	H	I	J	Total
Weekly Income (X)	850	700	100	750	5,000	80	420	2,500	400	360	11,160

$$\bar{X} = \frac{\Sigma X}{N} = \frac{11,160}{10} = ₹\ 1,116$$

Assumed Mean/Shortcut Method

$$\bar{X} = A + \frac{\Sigma d}{N}$$

where, d = (X – A); A = Assumed mean and N = Total number of observations.

For example:

The following table gives the daily income of tem workers in a factory. Find the arithmetic mean by Assumed Mean Method. (Take A = 250).

Workers	A	B	C	D	E	F	G	H	I	J
Daily Income (in ₹)	120	150	180	200	250	300	220	350	370	260

Solution: Computation of Arithmetic Mean by Assumed Mean Method.

Workers	Daily Income (₹) (X)	D = X – 250
A	120	–130
B	150	–100
C	180	–70
D	200	–50
E	250	0
F	300	+50
G	220	–30
H	350	+100
I	370	+120
J	260	+10
N = 10		–100

$$\bar{X} = A + \frac{\Sigma d}{N} = 250 + \left(\frac{-100}{10}\right) = 250 - 10 = ₹240$$

Thus, the average daily income of a worker is ₹ 240.

Step Deviation Method

$$\bar{X} = A + \frac{\Sigma d'}{N} \times c$$

where, $d' = \frac{(X-A)}{C}$;C is the common factor in d;

A = Assumed mean and N = Total number of observations.

For example:

The following data shows the weekly income of 10 families. Calculate the arithmetic mean by step deviation method and interpret the result. (Take A = 850).

Family	A	B	C	D	E	F	G	H	I	J
Weekly Income (in □)	850	700	100	750	5,000	80	420	2,500	400	360

Workers	Income (X)	d = X – 850	d' = (X – 850)/10
A	850	0	0
B	700	−150	−15
C	100	−750	−75
D	750	−100	−10
E	5000	+4150	+415
F	80	−770	−77
G	420	−430	−43
H	2500	+1650	+165
I	400	−450	−45
J	360	−490	−49
			+266

$$\bar{X} = A + \frac{\Sigma d'}{N} \times c = 850 + \frac{266}{10} \times 10 = 850 + 266 = ₹\ 1{,}116.$$

Interpretation: The average weekly income of a family is ₹ 1,116.

Discrete Series

Direct Method

$$\bar{X} = \frac{\Sigma fX}{\Sigma f}$$

where, ΣfX = Sum of all observations multiplied by their respective frequency and Σf = N = Total number of observations.

For example:

Plots in a housing colony come in only three sizes: 100 sq. metre, 200 sq. meters and 300 sq. metre and the number of plots are respectively 200, 50 and 10. Calculate the mean plot size in the housing colony by direct method.

Solution:

Computation of Arithmetic Mean by Direct Method.

Plot Size in sq. metre (X)	No. of Plots (f)	f X
100	200	20,000
200	50	10,000
300	10	3,000
	260	33,000

$$\bar{X} = \frac{\Sigma fX}{\Sigma f} = \frac{33000}{260} = 126.92$$

Therefore, the mean plot size in the housing colony is 126.92 sq. metre.

Assumed Mean/Shortcut Method

$$\bar{X} = A + \frac{\Sigma fX}{\Sigma f}$$

where, d = (X – A) ; A = Assumed mean and Σf = N = Total number of observations.

For example:

Plots in a housing colony come in only three sizes: 100 sq. meters and 300 sq. metre and the number of plots are respectively 200, 50 and 10. Calculate the mean plot size in the housing colony by assumed mean method (Take A = 200).

Solution:

Computation of Arithmetic Mean by Assumed Mean Method.

Plot Size in sq. metre (X)	No. of Plots (f)	d = X – 200	fd
100	200	−100	−20,000
200	50	0	0
300	10	+100	+1,000
			−19,000

Mean plot size,

$$\bar{X} = A + \frac{\Sigma fd}{\Sigma f} = 200 + \left(\frac{-19{,}000}{260}\right) = 200 - 73.08 = 126.92 \text{ sq.metre.}$$

Step Deviation Method

$$\bar{X} = A + \frac{\Sigma fd'}{\Sigma f} \times c$$

where, $d' = \frac{(X - A)}{C}$;

and C is the common factor in d.

A = Assumed mean and $\Sigma f = N$ = Total number of observations.

For example:

Plots in a housing colony come in only three sizes: 100 sq. metre, 200 sq. meters and 300 sq. metre and the number of plots are respectively 200, 50 and 10. Calculate the mean plot size in the housing colony by step deviation method. (Take A = 200).

Solution:

Computation of Arithmetic Mean by Step Deviation Method.

Plot Size in sq. metre (X)	No. of Plots (f)	d' = (X – 200)/100	fd'
100	200	−1	−200
200	50	0	0
300	10	+1	10
	260		−190

Mean Plot size $\bar{X} = A + \frac{\Sigma fd'}{\Sigma f} \times c = 200 + \left(\frac{-190}{260}\right) \times 100 = 200 - 73.08 = 126.92$ sq.metre

Continuous Series

Direct Method

$$\bar{X} = \frac{\Sigma fm}{\Sigma f}$$

where, Σfm = Sum of midpoints of classes multiplied by their respective class frequency.

$\Sigma f = N$ = Total number of observations.

For example:

Calculate average marks of the following:

Marks	0-10	10-20	20-30	0-40	40-50	50-60	60-70
No. of Students	5	12	15	25	8	3	2

Mark (X)	No. of students (g)	Mid value (m)	fm
(1)	(2)	(3)	(4) = (2) × (3)
0-10	5	5	25
10-20	12	15	180
20-30	15	25	375
30-40	25	35	875
40-50	8	45	360
50-60	3	55	165
60-70	2	65	130
	70		2,110

$$\bar{X} = \frac{\Sigma fm}{\Sigma f} = \frac{2{,}110}{70} = 30.14$$

Therefore, average marks of 70 students is 30.14.

Assumed Mean/Shortcut Method

$$\bar{X} = A + \frac{\Sigma fd}{\Sigma f}$$

where, d = (m – A) and m is the midpoint of the respective class.

A = Assumed mean and Σf = N = Total number of observations.

For example:

Calculate average marks of the following students by using assumed mean method: (Take A = 35).

Marks	0-10	10-20	20-30	0-40	40-50	50-60	60-70
No. of Students	5	12	15	25	8	3	2

Mark (X)	No. of students (f)	Mid value (m)	fd
(1)	(2)	(3)	(5) = (2) × (3)
0-10	5	5	−150
10-20	12	15	−240
20-30	15	25	−150
30-40	25	35	0
40-50	8	45	80
50-60	3	55	60
60-70	2	65	60
	70		−340

$$\bar{X} = A + \frac{\Sigma fd}{\Sigma f} = 35 + \left(\frac{-340}{70}\right) = 35 - 4.86 = 30.14$$

Therefore, Average marks of 70 students is 30.14.

Step Deviation Method

$$\bar{X} = A + \frac{\Sigma fd}{\Sigma f} \times c$$

where, $d' = \frac{(m - A)}{C}$; m is the midpoint of the respective class and C is the common factor in d. A = Assumed mean and Σf = N = Total number of observations.

For example:

Calculate average marks of the following students by using step deviation method: (Take A = 35).

Marks	0-10	10-20	20-30	0-40	40-50	50-60	60-70
No. of Students	5	12	15	25	8	3	2

Solution:

Computation of Average Marks for Exclusive Class Interval by Step deviation Method.

Mark (X)	No. of students (f)	Mid value (m)	d' = (m – 35)/10	fd'
(1)	(2)	(3)	(4)	(5) = (2) × (4)
0-10	5	5	−3	−15
10-20	12	15	−2	−24
20-30	15	25	−1	−15
30-40	25	35	0	0
40-50	8	45	1	8
50-60	3	55	2	6
60-70	2	65	3	6
	70			−34

Average marks of 70 students $\bar{X} = A + \frac{\Sigma fd'}{\Sigma f} \times c = 35 + \left(\frac{-34}{70}\right) \times 10 = 30.14$

Disadvantages of Mean

The mean has one main disadvantage: It is particularly susceptible to the influence of outliers. These are values that are unusual compared to the rest of the data set by being especially small or large in numerical value. For example, consider the wages of staff at a factory below:

Staff	1	2	3	4	5	6	7	8	9	10
Salary	15k	18k	16k	14k	15k	15k	12k	17k	90k	95k

The mean salary for these ten staff is $30.7k. However, inspecting the raw data suggests that this mean value might not be the best way to accurately reflect the typical salary of a worker, as most workers have salaries in the $12k to 18k range. The mean is being skewed by the two large salaries. Therefore, in this situation, we would like to have a better measure of central tendency.

Another time when we usually prefer the median over the mean (or mode) is when our data is skewed (i.e., the frequency distribution for our data is skewed). If we consider the normal distribution - as this is the most frequently assessed in statistics - when the data is perfectly normal, the mean, median and mode are identical. Moreover, they all represent the most typical value in the data set. However, as the data becomes skewed the mean loses its ability to provide the best central location for the data because the skewed data is dragging it away from the typical value. However, the median best retains this position and is not as strongly influenced by the skewed values.

Median and Mode

The median is the middle score for a set of data that has been arranged in order of magnitude. The median is less affected by outliers and skewed data. In order to calculate the median, suppose we have the data below:

65	55	89	56	35	14	56	55	87	45	92

We first need to rearrange that data into order of magnitude (smallest first):

14	35	45	55	55	56	56	65	87	89	92

Our median mark is the middle mark - in this case, 56 (highlighted in bold). It is the middle mark because there are 5 scores before it and 5 scores after it. This works fine when you have an odd number of scores, but what happens when you have an even number of scores? What if you had only 10 scores? Well, you simply have to take the middle two scores and average the result. So, if we look at the example below:

65	55	89	56	35	14	56	55	87	45

We again rearrange that data into order of magnitude (smallest first):

14	35	45	55	55	56	56	65	87	89

Only now we have to take the 5th and 6th score in our data set and average them to get a median of 55.5.

Mode

The mode is the most frequent score in our data set. On a histogram it represents the highest bar in a bar chart or histogram. You can, therefore, sometimes consider the mode as being the most popular option. An example of a mode is presented below:

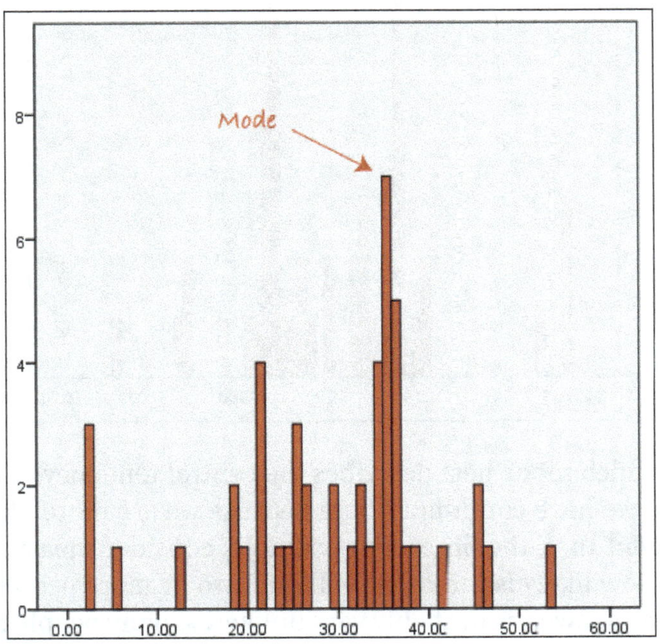

Normally, the mode is used for categorical data where we wish to know which is the most common category, as illustrated below:

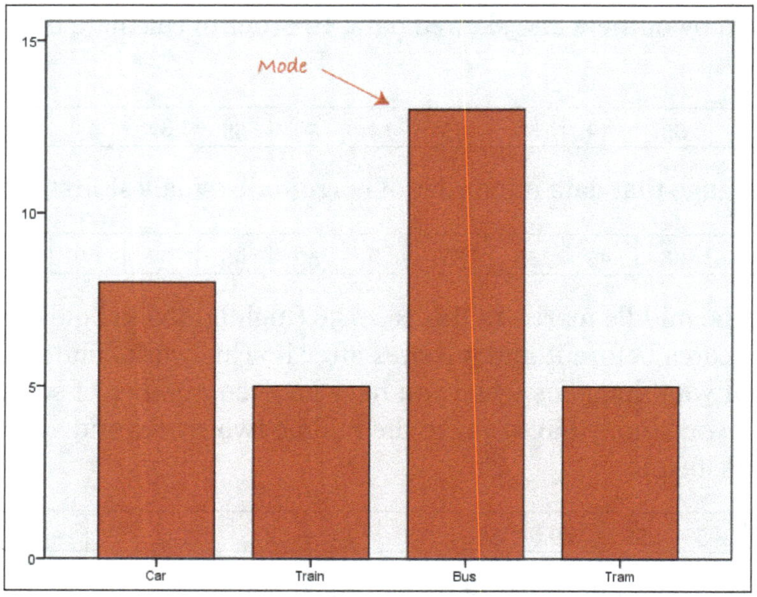

We can see above that the most common form of transport, in this particular data set, is the bus. However, one of the problems with the mode is that it is not unique, so it leaves us with problems when we have two or more values that share the highest frequency, such as below:

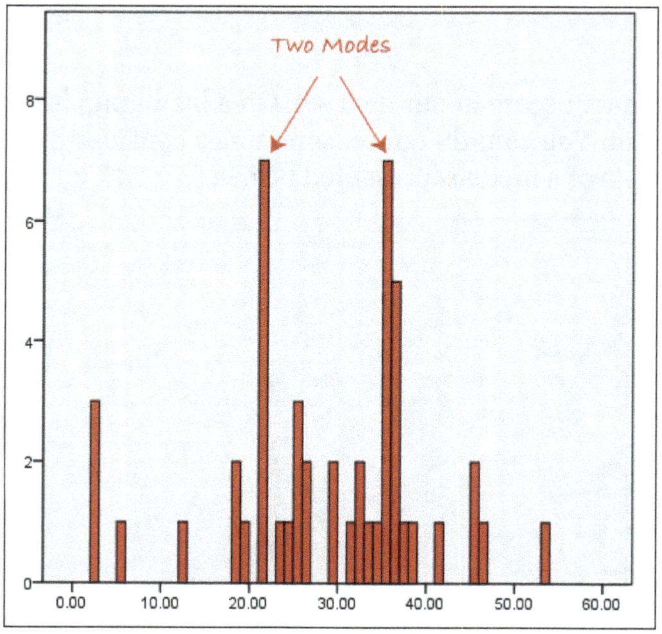

We are now stuck as to which mode best describes the central tendency of the data. This is particularly problematic when we have continuous data because we are more likely not to have any one value that is more frequent than the other. For example, consider measuring 30 peoples' weight (to the nearest 0.1 kg). How likely is it that we will find two or more people with exactly the same weight (e.g., 67.4 kg)? The answer, is probably very unlikely - many people might be close, but with such a small sample (30 people) and a large range of possible weights, you are unlikely to find two people with exactly the same weight; that is, to the nearest 0.1 kg. This is why the mode is very rarely used with continuous data.

Another problem with the mode is that it will not provide us with a very good measure of central tendency when the most common mark is far away from the rest of the data in the data set, as depicted in the diagram below:

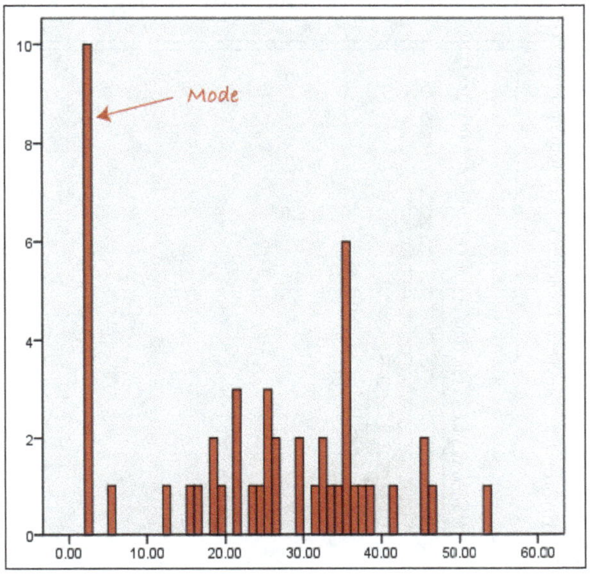

In the above diagram the mode has a value of 2. We can clearly see, however, that the mode is not representative of the data, which is mostly concentrated around the 20 to 30 value range. To use the mode to describe the central tendency of this data set would be misleading.

Skewed Distributions and the Mean and Median

We often test whether our data is normally distributed because this is a common assumption underlying many statistical tests. An example of a normally distributed set of data is presented below:

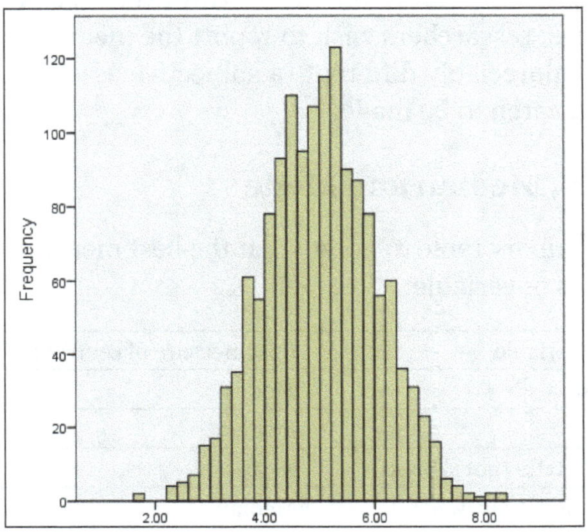

When you have a normally distributed sample you can legitimately use both the mean or the median as your measure of central tendency. In fact, in any symmetrical distribution the mean, median and mode are equal. However, in this situation, the mean is widely preferred as the best measure of central tendency because it is the measure that includes all the values in the data set for its calculation, and any change in any of the scores will affect the value of the mean. This is not the case with the median or mode.

However, when our data is skewed, for example, as with the right-skewed data set below:

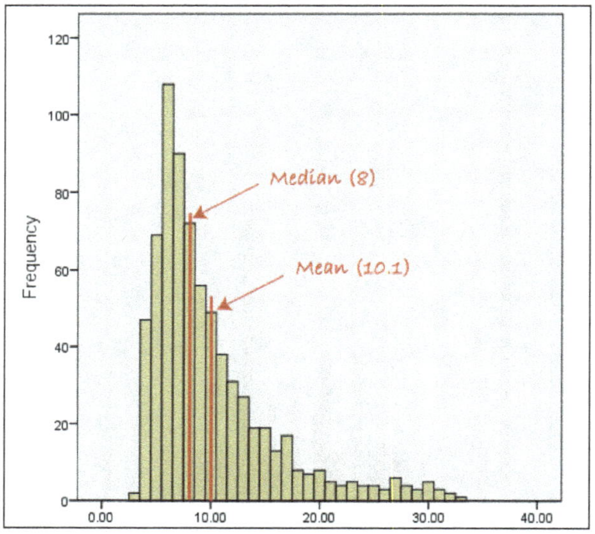

We find that the mean is being dragged in the direct of the skew. In these situations, the median is generally considered to be the best representative of the central location of the data. The more skewed the distribution, the greater the difference between the median and mean, and the greater emphasis should be placed on using the median as opposed to the mean. A classic example of the above right-skewed distribution is income (salary), where higher-earners provide a false representation of the typical income if expressed as a mean and not a median.

If dealing with a normal distribution, and tests of normality show that the data is non-normal, it is customary to use the median instead of the mean. However, this is more a rule of thumb than a strict guideline. Sometimes, researchers wish to report the mean of a skewed distribution if the median and mean are not appreciably different (a subjective assessment), and if it allows easier comparisons to previous research to be made.

When to use the Mean, Median and Mode

Please use the following summary table to know what the best measure of central tendency is with respect to the different types of variable:

Type of Variable	Best measure of central tendency
Nominal	Mode
Ordinal	Median
Interval/Ratio (not skewed)	Mean
Interval/Ratio (skewed)	Median

Application of Statistics in Environmental Science

In inferential statistics, concept of probability is important for studying the uncertainties in environment. For example, whether it will rain or not tomorrow can be best inferred by using probability. Several theoretical probability distribution viz. Bernoulli distribution, binomial distribution, Poisson distribution etc. are useful for modelling probability distribution of real environmental data. For example decisions such as coin tossing, rain/no rain, yes/no etc. are explained by Bernoulli variable since their outcomes are binary. Also if we are interested in counting the number of times flood occurs out of a total number of floods occurred. The probability distribution of such variable is given by Binomial distribution since we are to count the number of times flood occur (X), a Bernoulli event, with probability p out of a total i.e. out of n trials. Also, if we do not know the total number of occurrence of floods, but know the mean of the occurrence of flood then the distribution is modelled by Poisson distribution. Statistical tools such as estimation i.e estimating population parameters from sample values, hypothesis testing etc. plays a vital role in analysing environmental data. Some of the frequently used teat statistic in atmospheric and environmental science is Z test, t test, F test etc. Another statistical approach is time series analysis, which study environmental quantities with respect to time. For example, monthly/yearly mean temperature, rainfall, humidity etc is best studied by time series.

References

- Discover-data-blog-series, data-types-101: chi2innovations.com, Retrieved 13, March 2020
- Environmental-data, ecodata, schedule, teaching: umass.edu, Retrieved 25, June 2020
- Graphic-presentation-of-data, descriptive-statistics, business-economics-cs: toppr.com, Retrieved 08, February 2020
- Relation-between-am-gm-hm, maths: vedantu.com, 29, May 2020
- Measures-central-tendency-mean-mode-median, statistical-guides: laerd.com, 14, August 2020

2

Probability: An Integrated Study

The branch of mathematics which deals with the likelihood of the occurrence of an event is known as probability. A few of the different types of events which are studied within probability are simple event, compound event, exhaustive event and dependent event. All the diverse aspects of these types of events in probability have been carefully analyzed in this chapter.

In environmental science and many other subjects there is certainly no shortage of uncertainty, for example, in our knowledge about whether it will rain or not tomorrow. Uncertainty about such events arises naturally from errors and gaps in measurements, incomplete and incorrect knowledge of the underlying mechanisms, and also from the overall complexity of all the possible interactions in real-world systems. We try to describe this uncertainty qualitatively by using words such as "likely", "probably", "chance", etc.. However, to make progress scientifically it is necessary to use a more quantitative definition of uncertainty. In 1812, the French mathematician Pierre Simon Laplace defined the word "probability" to mean a number lying between 0 and 1 that measures the amount of certainty for an event to occur. A probability of 1 means the event is completely certain to occur, whereas a probability of 0 means that the event will certainly never occur. Probability is essential for understanding how samples of data can be drawn from the underlying population, and for making inferences about this population based on sample statistics.

Intuitive Set Theory

We need a few notions from set theory before we jump into probability theory. In doing so we will use intuitive or "naive" definitions. This intuitive approach provides good mnemonics and is sufficient for our purposes but soon runs into problems for more advanced applications.

- Set: A set is a collection of elements. Sets are commonly represented using curly brackets containing a collection of elements separated by commas. For example:

  $$A = \{1, 2, 3\}$$

  tells us that A is a set whose elements are the first 3 natural numbers. Sets can also be represented using a rule that identifies the elements of the set. The prototypical notation is as follows:

  $$\{x : x \text{ follows a rule}\}$$

  For example,

  $$\{x : x \text{ is a natural number and } x \text{ is smaller than } 4\}$$

- Outcome Space: The outcome space is a set whose elements are all the possible basic outcomes of an experiment. The sample space is also called sample space, reference set, and universal set and it is commonly represented with the capital Greek letter "omega", Ω. We call the elements of the sample space "outcomes" and represent them symbolically with the small Greek letter "omega", ω.

Example: If we roll a die, the outcome space could be:

$$\Omega = \{1, 2, 3, 4, 5, 6\}$$

In this case the symbol ω could be used to represent either 1, 2, 3, 4, 5 or 6.

Example: If we toss a coin twice, we can observe 1 of 4 outcomes: (Heads, Heads), (Heads, Tails), (Tails, Heads), (Tails, Tails). In this case we could use the following outcome space,

$$\Omega = \{(H, H),(H, T),(T, H),(T, T)\}$$

and the symbol ω could be used to represent either (H, H), or (H, T), or (T, H), or (T, T). Note how in this case each basic outcome contains 2 elements. If we toss a coin n times each basic outcome ω would contain n elements.

- Singletons: A singleton is a set with a single element. For example the set {4} is a singleton, since it only has one element. On the other hand 4 is not a singleton since it is an element not a set.
- Element inclusion: We use the symbol $\in$ to represent element inclusion. The expression $\omega \in$ A tells us that ω is an element of the set A. The expression $\omega \notin$ A tells us that ω is not an element of the set A. For example, $1 \in \{1, 2\}$ is true since 1 is an element of the set {1, 2}. The expression $\{1\} \in \{\{1\}, 2\}$ is also true since the singleton {1} is an element of the set {{1}, 2}. The expression $\{1\} \notin \{1, 2\}$ is also true, since the set {1} is not an element of the set {1, 2}.
- Set inclusion: We say that the set A is included in the set B or is a subset of B if all the elements of A are also elements of B. We represent set inclusion with the symbol $\subset$. The expression $A \subset B$ tells us that both A and B are sets and that all the elements of A are also elements of B. For example the expression $\{1\} \subset \{1, 2\}$ is true since all the elements of the set {1} are in the set {1, 2}. On the other hand $1 \subset \{1, 2\}$ is not true since 1 is an element, not a set.
- Set equality: Two sets A and B are equal if all elements of A belong to B and all elements of B belong to A. In other words, if $A \subset B$ and $B \subset A$. For example the sets {1, 2, 3} and {3, 1, 1, 2, 1} are equal.
- Set Operations: There are 3 basic set operations:
    - Union: The union of two sets A and B is another set that includes all elements of A and all elements of B. We represent the union operator with this symbol $\cup$,

      For example, if A = {1, 3, 5} and B = {2, 3, 4}, then $A \cup B = \{1, 2, 3, 4, 5\}$. More generally,

      $A \cup B = \{\omega : \omega \in A \text{ or } \omega \in B\}$.

In other words, the set $A \cup B$ is the set of elements with the property that they either belong to the set A or to the set B.

- Intersection: The intersection of two sets A and B is another set C such that all elements in C belong to A and to B. The intersection operator is symbolized as $\cap$. If $A = \{1, 3, 5\}$ and $B = \{2, 3, 4\}$ then $A \cap B = \{3\}$. More generally,

$$A \cap B = \{\omega : \omega \in A \text{ and } \omega \in B\}$$

- Complementation: The complement of a set A with respect to a reference set Ω is the set of all elements of Ω which do not belong to A. The complement of A is represented as A^c. For example, if the universal set is {1, 2, 3, 4, 5, 6} then the complement of {1, 3, 5} is {2, 4, 6}. More generally,

$$A^c = \{\omega : \omega \in \Omega \text{ and } \omega \notin A\}$$

- Empty set: The empty set is a set with no elements. We represent the null set with the symbol $\emptyset$. Note $\Omega^c = \emptyset$, $\emptyset^c = \Omega$, and for any set A,

$$A \cup \emptyset = A$$
$$A \cap \emptyset = \emptyset$$

- Disjoint sets: Two sets are disjoint if they have no elements in common, i.e., their intersection is the empty set. For example, the sets {1, 2} and {1} are not disjoint since they have an element in common.

- Collections: A collection of sets is a set of sets, i.e., a set whose elements are sets. For example, if A and B are the sets defined above, the set {A, B} is a collection of sets.

- Power set: The power set of a set A is the a collection of all possible sets of A. We represent it as $\mathfrak{B}(A)$. For example, if $A = \{1, 2, 3\}$ then,

$$\mathfrak{B}(A) = \{\emptyset, \{1\}, \{2\}, \{3\}, \{1, 2\}, \{1, 3\}, \{2, 3\}, A\}$$

  Note that 1 is not an element of $\mathfrak{B}(A)$ but {1} is. This is because 1 is an element of A, not a set of A.

- Collections closed under set operations: A collection of sets is closed under set operations if any set operation on the sets in the collection results in another set which still is in the collection. If $A = \{1, 3, 5\}$ and $B = \{2, 3, 4\}$, the collection $C = \{A, B\}$ is not closed because the set $A \cap B = \{3\}$ does not belong to the collection. The collection $C = \{\Omega, \emptyset\}$ is closed under set operations, all set operations on elements of C produce another set that belongs to C. The power set of a set is always a closed collection.

- Sigma algebra: A sigma algebra is a collection of sets which is closed when set operations are applied to its members a countable number of times. The power set of a set is always a sigma algebra.

- Natural numbers: We use the symbol $\mathbb{N}$ to represent the natural numbers, i.e., {1, 2, 3, . . .}.

One important property of the natural numbers is that if $x \in \mathbb{N}$ then $x + 1 \in \mathbb{N}$.

- Integers: We use the symbol Z to represent the set of integers, i.e., {. . . , −3, −2, −1, 0, 1, 2, 3, . . .}. Note $\mathbb{N} \subset \mathbb{Z}$. One important property of the natural numbers is that if $x \in \mathbb{Z}$ then $x + 1 \in \mathbb{Z}$ and $x - 1 \in \mathbb{Z}$.
- Real numbers: We use the symbol R to represent the real numbers, i.e., numbers that may have an infinite number of decimals. For example, 1, 2.35, −4/123, $\sqrt{2}$, and π, are real numbers. Note $\mathbb{N} \subset \mathbb{Z} \subset \mathbb{R}$.
- Cardinality of sets:
    - We say that a set is finite if it can be put in one-to-one correspondence with a set of the form {1, 2, . . . , n}, where n is a fixed natural number.
    - We say that a set is infinite countable if it can be put in one-to-one correspondence with the natural numbers.
    - We say that a set is countable if it is either finite or infinite countable.
    - We say that a set is infinite uncountable if it has a subset that can be put in one-to-one correspondence with the natural numbers, but the set itself cannot be put in such a correspondence. This includes sets that can be put in one-to-one correspondence with the real numbers.

Understanding Probability

If, one is interested in particular results among all possible results of a random experiment, these particular results can be obtained in terms of events.

"Event is a particular performance of a random experiment".

List of some examples of event:

- Suppose a random experiment is tossing a coin two times.
    - Sample space: {HH, HT, TH, TT}.
    - Possible event: getting head both times, getting tail both times, getting one head and tail of a coin.
- Suppose rolling a die two times is a random Experiment.
    - Sample space: {(1,1), (1,2),.........,(1,6), (2,1), (2,2), ...,(2,6), (3,1), (3,2),.......,(3,6), (4,1), (4,2), ...,(4,6), (5,1), (5,2),.......,(5,6), (6,1), (6,2),.....,(6,6)}.

    - Possible event: Getting same number both times , getting odd number at first time and even number at second time, getting odd number at first time, getting sum of both is even, getting sum of both time is odd , many more event may occur.
- Selecting a card from a deck of cards is an random experiment.
    - Sample space: {1, 2, ...,10, A, K, Q, J}.
    - Possible event: Getting a spade, getting a club, getting a heart , getting a diamond, getting a face card, getting a king, getting a queen , many more event may occur respect to sample space of random experiment.

Types of Events

There are many types of event according to the happening of event and outcome.

Simple Event

Suppose a particular event happens among all possible events. Result of particular event is at most one, then event is called simple event.

"An event that shows exactly one outcome of a random experiment is a simple event".

Example:

1. Suppose a quality analyst analysis two processes' product in a random experiment and classify the process product as defective product and non-defective product.

- Sample space: {nn, nd, dn, dd}.

  where d is for defective product and n is for non-defective product.

  Let E is an event.

  Event E: Neither product is defective.

  E ={nn}; as E has only one outcome, E is a simple event.

2. Tossing a coin two times in a random experiment.

- Sample space: {HH, HT, TH, TT}; where H: represents head, T: represents tail.

  Let E be an event: getting head both time.

  E = {HH}; E has only one outcome, E is a simple event.

Compound Event

Suppose a particular event happens among all possible events. Result of particular event is at least one outcome that event is called a compound event.

"A event shows more than one outcome of a random experiment is a compound event".

Example:

1. Suppose a quality analyst analysis two processes' product quality in an random experiment and classify the process product as defective product and non-defective product.

- Sample space : {nn, nd, dn, dd}

  where: d is for defective product and n is for non-defective product.

  Let E is an event.

  Event E : at least one product is defective.

  E: {nd, dn, dd}; E has more than one outcome. So, E is a compound event.

2. Tossing a coin two times is a random experiment.

- Sample space: {HH, HT, TH, TT}; where H: represents head, T: represents tail.

  Let E be an event getting head at least one time.

  E: {HH, HT, TH}; E has more than one outcome. So, E is a compound event.

Exhaustive Event

All possible results of an experiment is termed as an exhaustive event.

"Exhaustive event is the total outcome of a random experiment".

List of some examples are:

- Tossing of one coin is a random experiment, exhaustive cases are 2.
- Throwing a dice, exhaustive cases are 6.
- Throwing two dices, exhaustive cases are 6^2.
- Tossing two coins, exhaustive cases are 2^2.
- Selecting a card, exhaustive cases are 5^2.

Dependent Event

Two particular events E and F occurred in a trail. Result of one event are affecting the result of second event. Event may be either E or F, these events will be dependent event.

"Event is said to be dependent event if happening of one event in one trail is affected by happening of similar event in next trail".

Examples:

- Drawing a card from a pack of card followed by one after another draw is a random experiment. Let us consider E is an event of drawing black card. If first drawn card is not replaced in the deck of card while second card is drawn, then the second time drawn card will be dependent on the first drawn card. After this total number of cards are decreasing followed by one after another draw. So, E is a dependent event.

- Let us consider a bag with five marbles, four are red and one is blue. If a marble is selected at random from the bag and not replaced again into the bag, the chance of occurrence of the blue ball will keep increasing with each consecutive draw.

Favorable Event

Outcomes of a particular event are favorable to that events. These outcomes of events are considered as favorable event. "Favorable event is the number of outcome favor to an event".

Examples:

- Tossing two coins is a random experiment and getting head on both coins is an event. Exhaustive event is $2^2 = 4$ and favorable event is 1 (i.e. head and head).
- Throwing two dice is an random experiment and getting same number on both side is an event, favorable event is 6 {(1,1), (2,2), (3,3), (4,4), (5,5), (6,6)}.

Mutually Exclusive Event

"Events can't occur simultaneously in same trial are mutually exclusive events".

Suppose A and B are two events. If A and B are mutually exclusive event, their venn diagram is given in figure.

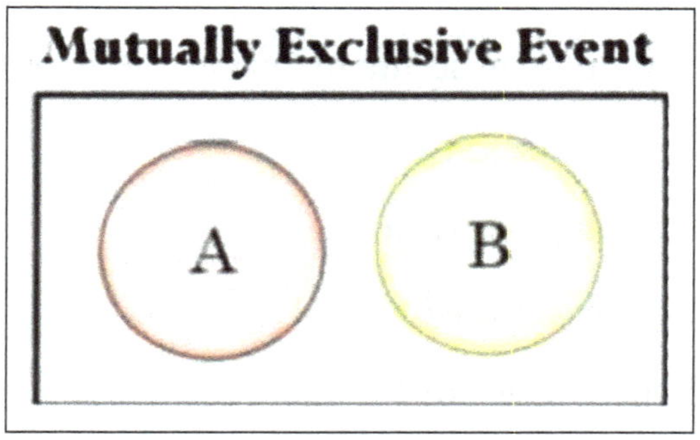

Examples:

- Head and tail are two outcomes of random experiment of a coin. Suppose A and B are two events of getting head and getting tail respectively. Both event cannot occur in a single trail, so A and B are mutually exclusive event.
- 1, 2, 3, 4, 5, 6 are the possible six outcome of random experiment in throwing a dice. All of these outcomes cannot occur in a single trail, so they are all mutually exclusive event.
- A switch cannot be on/off at same time. Switching a switch button is a random experiment that has two possible outcomes on and off that switch is on while the second one shows switch is off. Both event cannot occur at same time, so A and B are mutually exclusive event.

Independent Event

Two particular event E and F are occurred in a trial. Result of one event is not affecting the result of second event. Event may be either E or F, so these events will be independent event.

"Event is said to be independent event if happening of one event in one trail is not affected by happening of similar event in next trial".

Examples:

- Throwing a dice is a random experiment, if one get six in first throw as well as in second throw. One can see that both throws of a dice are not affecting the outcome of each other, so these are independent event.
- Tossing a coin is a random experiment, if one get head on first toss as well as second toss. One can see that both tosses of a coin are not affecting the outcome of each other, so these are independent event.
- Drawing a card from a pack of card with replacement is independent event. Every time total number of cards remain same.

Equally Likely Events

All possible event of a random experiment have same numerical value of probability of occurring events. These events are known as equally likely events with equal probability of occurrence.

"Equally likely events shows equal probability for all events. There is no preference for any event, all have equal preference".

Example:

1. Tossing a coin is a random experiment. Possible outcomes are head and tail. Possible event are getting head and getting tail. One has total possible outcome as 2.

Probability of getting head is ½.

Probability of getting tail is ½.

Both have equal probability and equal preference. So, events are equally likely events.

2. Throwing a die is a random experiment with six possible outcomes and events. All possible event have equal probability i.e. 1/6. So all events have equal preference to happening, known as equally likely events.

Impossible Event

If an event under a random experiment cannot occur, that event is known as an impossible event.

"Numerical measure of an event is zero then the event is impossible event".

For an impossible event E,

$$(E) = 0.$$

Example:

If tossing a dice is a random experiment. E is an event of outcome greater than 6.We know that dice have six faces only so there is no chance of happening this event. Hence event E is an impossible event.

Certain Event

If there is an absolute surety on the chance of occurrence of an event, event is known as certain event. Suppose if it is Thursday, the probability that tomorrow is Friday is certain event with one probability.

"Numerical measure of an event is one, then event is certain event".

For a sure and certain event E:

$$(E) = 1.$$

Example:

Hitting a dart is a random experiment and throwing a dart at center is an event on the dartboard. Imagine that this is only one thing in the universe and dart hit at the center, there is only one chance to throw then this event is sure event.

Complementary Event

$$P(A)+P(A^C)=1.$$

Example:

- Tossing a coin is a random experiment and we are interested in getting head. If head appears then it is a happening event, getting tail is non-happening event and complementary event.
- Rolling a dice is a random experiment and getting 6 on dice is event. Getting value other then 6 or not getting 6 is non-happening event and complementary event.

Probability Approaches

There are three ways to assign probabilities to events: classical approach, relative-frequency approach, subjective approach.

Classical Approach

If an experiment has n simple outcomes, this method would assign a probability of 1/n to each outcome. In other words, each outcome is assumed to have an equal probability of occurrence.

This method is also called the axiomatic approach.

Example:

1. Roll of a Die:

$$S = \{1, 2, \cdots, 6\}$$

Probabilities: Each simple event has a 1/6 chance of occurring.

2. Two Rolls of a Die:

$$S = \{(1, 1), (1, 2), \cdots, (6, 6)\}$$

Assumption: The two rolls are "independent."

Probabilities: Each simple event has a $(1/6) \cdot (1/6) = 1/36$ chance of occurring.

Relative-Frequency Approach

Probabilities are assigned on the basis of experimentation or historical data.

Formally, Let A be an event of interest, and assume that you have performed the same experiment n times so that n is the number of times A could have occurred. Further, let n_A be the number of times that A did occur. Now, consider the relative frequency n_A/n. Then, in this method, we "attempt" to define P(A) as:

$$P(A)+\lim_{n\to\infty}\frac{n_A}{n}.$$

The above can only be viewed as an attempt because it is not physically feasible to repeat an experiment an infinite number of times. Another important issue with this definition is that two sets of n experiments will typically result in two different ratios. However, we expect the discrepancy to converge to 0 for large n. Hence, for large n, the ratio n_A/n may be taken as a reasonable approximation for P(A).

Example:

1. Roll of a Die:

$$S = \{1, 2, \cdots, 6\}$$

Probabilities: Roll the given die 100 times (say) and suppose the number of times the outcome 1 is observed is 15. Thus, $A = \{1\}$, $n_A = 15$, and $n = 100$. Therefore, we say that P(A) is approximately equal to $15/100 = 0.15$.

2. Computer Sales: A computer store tracks the daily sales of desktop computers in the past 30 days.

The resulting data is:

Desktops Sold	No. of Days
0	1
1	2

2	10
3	12
4	5
5 or more	0

The approximate probabilities are:

Desktops Sold	No. of Days	Probability
0	1	1/30 = 0.03
1	2	2/30 = 0.07
2	10	10/30 = 0.33
3	12	12/30 = 0.40
4	5	5/30 = 0.17
5 or more	0	0

Thus, for example, there is a 40% chance that the store will sell 3 desktops on any given day.

Subjective Approach

In the subjective approach, we define probability as the degree of belief that we hold in the occurrence of an event. Thus, judgment is used as the basis for assigning probabilities.

Notice that the classical approach of assigning equal probabilities to simple events is, in fact, also based on judgment.

What is somewhat different here is that the use of the subjective approach is usually limited to experiments that are unrepeatable.

Example:

1. Horse Race: Consider a horse race with 8 horses running. What is the probability for a particular horse to win? Is it reasonable to assume that the probability is 1/8? Note that we can't apply the relative-frequen-cy approach.

People regularly place bets on the outcomes of such "onetime" experiments based on their judgment as to how likely it is for a particular horse to win. Indeed, having different judgments is what makes betting possible.

2. Stock Price: What is the probability for a particular stock to go up tomorrow? Again, this "experiment" can't be repeated, and we can't apply the relative-frequency approach.

Sophisticated models (that rely on past data) are often used to make such predictions, as blindly following ill-founded judgments is often dangerous.

Axioms of Probability

Numerical measure of uncertainty to respective event is called probability. For accurate and best result of an event as much as possible probability must follow some axioms. Axioms of probability are given as:

Axioms of Non-negativity

"Probability of an event must be non-negative".

Suppose E is any event of a random experiment, probability of E is:

$$P(E) \geq 0$$

Probability lies between one and zero, probability must be non-negative but less than one.

$$0 \leq P(E) \leq 1.$$

Axioms of Certainty

"Probability sum of all possible outcomes of a random experiment must be one".

Suppose E is an event of a random experiment, probability sum of all possible event is:

$$P(E) = 1.$$

If probability of getting head at first time of tossing a coin is 0.5 and probability of getting tail on first time of tossing a coin is 0.5. Total of getting head or tail is 1.

Axioms of Additivity

"Probability of union of all events is equal to the sum of the probability of all distinct event".

$$P\left(\bigcup_{i=1}^{n} A_i\right) = \sum_{i=1}^{n} P(A_i)$$

where A_i are mutually exclusive and exhaustive events ∀ i=1, 2, 3, 4, ...n.

Permutations and Combinations

General Counting Principle

If any event can occur in m ways and after it happens in any one of these ways, a second event can occur in n ways, then both the events together can occur in m × n ways.

Example:

1. How many multiples of 5 are there from 10 to 95?

Solution: As you know, multiples of 5 are integers having 0 or 5 in the digit to the extreme right (i.e. the unit's place).

The first digit from the right can be chosen in 2 ways.

The second digit can be any one of 1,2,3,4,5,6,7,8,9.

i.e. There are 9 choices for the second digit.

Thus, there are 2×9 =18 multiples of 5 from 10 to 95.

2. In a city, the bus route numbers consist of a natural number less than 100, followed by one of the letters A, B, C, D, E and F. How many different bus routes are possible?

Solution: The number can be any one of the natural numbers from 1 to 99. There are 99 choices for the number.

The letter can be chosen in 6 ways.

∴ Number of possible bus routes are 99 × 6 = 594.

3. There are 3 questions in a question paper. If the questions have 4, 3 and 2 solutionsvely, find the total number of solutions.

Solution: Here question 1 has 4 solutions, question 2 has 3 solutions and question 3 has 2 solutions.

∴ By the multiplication (counting) rule, total number of solutions = 4 × 3 × 2 = 24.

4. Consider the word ROTOR. Whichever way you read it, from left to right or from right to left, you get the same word. Such a word is known as palindrome. Find the maximum possible number of 5-letter palindromes.

Solution: The first letter from the right can be chosen in 26 ways because there are 26 alphabets.

Having chosen this, the second letter can be chosen in 26 ways.

∴ The first two letters can chosen in 26×26 = 676 ways.

Having chosen the first two letters, the third letter can be chosen in 26 ways.

∴ All the three letters can be chosen in 676×26 =17576 ways.

It implies that the maximum possible number of five letter palindromes is 17576 because the fourth letter is the same as the second letter and the fifth letter is the same as the first letter.

In Example we found the maximum possible number of five letter palindromes. There cannot be more than 17576. But this does not mean that there are 17576 palindromes. Because some of the choices like CCCCC may not be meaningful words in the English language.

5. How many 3-digit numbers can be formed with the digits 1,4,7,8 and 9 if the digits are not repeated.

Solution: Three digit number will have unit's, ten's and hundred's place.

Out of 5 given digits any one can take the unit's place.

1. This can be done in 5 ways.

- After filling the unit's place, any of the four remaining digits can take the ten's place.

2. This can be done in 4 ways.

- After filling in ten's place, hundred's place can be filled from any of the three remaining digits.

3. This can be done in 3 ways.

- By counting principle, the number of 3 digit numbers = $5 \times 4 \times 3 = 60$.

Let us now state the General Counting Principle.

If there are n events and if the first event can occur in m_1 ways, the second event can occur in m_2 ways after the first event has occured, the third event can occur in m_3 ways after the second event has ocurred, and so on, then all the n events can occur in $m_1 \times m_2 \times \ldots \times m_{n-1} \times m_n$ ways.

6. Suppose you can travel from a place A to a place B by 3 buses, from place B to place C by 4 buses, from place C to place D by 2 buses and from place D to place E by 3 buses. In how many ways can you travel from A to E?

Solution:

- The bus from A to B can be selected in 3 ways.
- The bus from B to C can be selected in 4 ways.
- The bus from C to D can be selected in 2 ways.
- The bus from D to E can be selected in 3 ways.

So, by the General Counting Principle, one can travel from A to E in $3 \times 4 \times 2 \times 3$ ways = 72 ways.

Permutations

Suppose you want to arrange your books on a shelf. If you have only one book, there is only one way of arranging it. Suppose you have two books, one of History and one of Geography.

You can arrange the Geography and History books in two ways. Geography book first and the History book next, GH or History book first and Geography book next; HG. In other words, there are two arrangements of the two books.

Now, suppose you want to add a Mathematics book also to the shelf. After arranging History and Geography books in one of the two ways, say GH, you can put Mathematics book in one of the following ways: MGH, GMH or GHM. Similarly, corresponding to HG, you have three other ways of arranging the books. So, by the Counting Principle, you can arrange Mathematics, Geography and History books in 3×2 ways = 6 ways.

By permutation we mean an arrangement of objects in a particular order. In the above example, we were discussing the number of permutations of one book or two books.

In general, if you want to find the number of permutations of n objects n ≥1, how can you do it? Let us see if we can find an answer to this.

Similar to what we saw in the case of books, there is one permutation of 1 object, 2 × 1 permutations of two objects and 3 × 2 × 1 permutations of 3 objects. It may be that, there are n × (n –1) × (n – 2) ×...× 2 × 1 permutations of n objects. In fact, it is so, as you will see when we prove the following result.

Theorem: The total number of permutations of n objects is n (n – 1) ....2.1.

Proof: We have to find the number of possible arrangements of n different objects.

The first place in an arrangement can be filled in n different ways. Once it has been done, the second place can be filled by any of the remaining (n–1) objects and so this can be done in (n–1) ways. Similarly, once the first two places have been filled, the third can be filled in (n–2) ways and so on. The last place in the arrangement can be filled only in one way, because in this case we are left with only one object.

Using the counting principle, the total number of arrangements of n different objects is:

$$n(n-1)(n-2)........2.1.$$

The product n (n – 1) ... 2.1 occurs so often in Mathematics that it deserves a name and notation. It is usually denoted by n! (or by n read as n factorial).

$$n! = n(n-1)\ldots 3.2.1$$

Here is an example to help you familiarise yourself with this notation.

Example: Evaluate (a) 3! (b) 2! + 4! (c) 2!×3!

Solution:

(a) $3! = 3 \times 2 \times 1 = 6$

(b) $2! = 2 \times 1 = 2$

$4! = 4 \times 3 \times 2 \times 1 = 24$

Therefore, $2! + 4! = 2 + 24 = 26$

(c) 2! 3! 2 6 12

Notice that n! satisfies the relation,

$$n! = n \times (n-1)\ldots$$

This is because, $n(n-1)! = n[(n-1).(n-2)\ldots 2.1]$,

$$= n.(n-1).(n-2)\ldots 2.1 = n!$$

Of course, the above relation is valid only for n ≥ 2 because 0! has not been defined so far. Let us see if we can define 0! to be consistent with the relation. In fact, if we define,

$$0! = 1\ldots$$

then the relation holds for n = 1 also.

Example: Suppose you want to arrange your English, Hindi, Mathematics, History, Geography and Science books on a shelf. In how many ways can you do it?

Solution: We have to arrange 6 books.

The number of permutations of n objects is $n! = n.(n-1).(n-2) \ldots 2.1$

Here $n = 6$ and therefore, number of permutations is $6.5.4.3.2.1 = 720$

Permutation of R Objects out of N Objects

Suppose you have five story books and you want to distribute one each to Asha, Akhtar and Jasvinder. In how many ways can you do it? You can give any one of the five books to Asha and after that you can give any one of the remaining four books to Akhtar. After that, you can give one of the remaining three books to Jasvinder. So, by the Counting Principle, you can distribute the books in $5 \times 4 \times 3$ ie. 60 ways.

More generally, suppose you have to arranger objects out of n objects. In how many ways can you do it? Let us view this in the following way. Suppose you haven objects and you have to arrange r of these in r boxes, one object in each box.

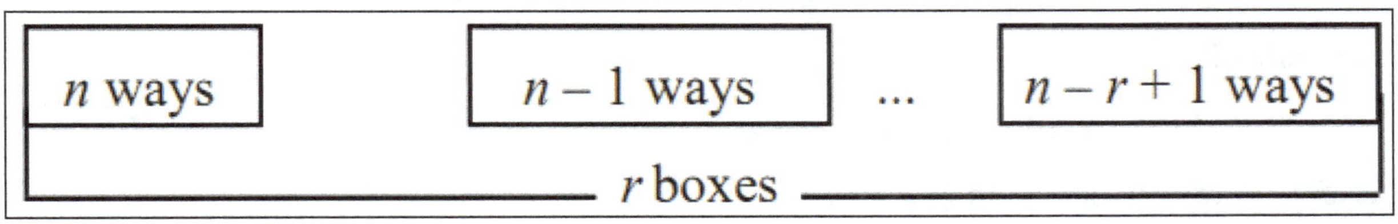

Suppose there is one box. $r = 1$. You can put any of the n objects in it and this can be done in n ways. Suppose there are two boxes. $r = 2$. You can put any of the objects in the first box and after that the second box can be filled with any of the remaining $n - 1$ objects. So, by the counting principle, the two boxes can be filled inn $(n - 1)$ ways. Similarly, 3 boxes can be filled in $n(n - 1)(n - 2)$ ways.

In general, we have the following theorem.

Theorem: The number of permutations of r objects out of n objects is:

$$n(n-1)\ldots(n-r+1).$$

The number of permutations of r objects out of n objects is usually denoted by ${}^nP_r$. Thus,

$${}^nP_r = n(n-1)(n-2)\ldots(n-r+1)$$

Proof: Suppose we have to arrange r objects out of n different objects. In fact it is equivalent to filling r places, each with one of the objects out of the given n objects.

The first place can be filled in n different ways. Once this has been done, the second place can be filled by any one of the remaining $(n-1)$ objects, in $(n-1)$ ways. Similarly, the third place can be filled in $(n - 2)$ ways and so on. The last place, the rth place can be filled in $[n-(r-1)]$ i.e. $(n-r+1)$ different ways.

Using the Counting Principle, we get the required number of arrangements of r out of n objects is n (n −1)(n − 2)............(n − r +1).

Example: Evaluate : $(a)\ {}^4P_2\ (b)\ {}^6P_3\ (c)\ \frac{{}^4P_3}{{}^3P_2}\ (d)\ {}^6P_3 \times {}^5P_2$

Solution:

$$(a)\ {}^4P_2 = 4(4-1) = 4\times3 = 12$$
$$(b)\ {}^6P_3 = 6(6-1)(6-2) = 6\times5\times4 = 120$$
$$(c)\ \frac{{}^4P_3}{{}^3P_2} = \frac{4(4-1)(4-2)}{3(3-1)} = \frac{4\times3\times2}{3\times2} = 4$$
$$(d)\ {}^6P_3 \times {}^5P_2 = 6(6-1)(6-2)\times5(5-1)$$
$$= 6\times5\times4\times5\times4 = 2400$$

Example: If you have 6 New Year greeting cards and you want to send them to 4 of your friends, in how many ways can this be done?

Solution: We have to find number of permutations of 4 objects out of 6 objects.

This number is:

$${}^6P_4 = 6(6-1)(6-2)(6-3) = 6.5.4.3 = 360$$

Therefore, cards can be sent in 360 ways.

Consider the formula for ${}^nP_r$, namely, ${}^nP_r$ = n (n − 1) ... (n − r + 1). This can be obtained by removing the terms n − r, n − r − 1,...,2, 1 from the product for n!. The product of these terms is (n − r) (n − r − 1) ...2.1, i.e., (n − r)!.

Now,

$$\frac{n!}{(n-r)!} = \frac{n(n-1)(n-2)...(n-r+1)(n-r)...2.1}{(n-r)(n-r-1)...2.1}$$
$$= n(n-1)(n-2)...(n-r+1)$$
$$= {}^nP_r$$

So, using the factorial notation, this formula can be written as follows:

$${}^nP_r = \frac{n!}{(n-r)!}$$

Example: Find the value of ${}^nP_r$.

Solution: Here r = 0. Using relation 7.5, we get,

$${}^nP_r = \frac{n!}{n!} = 1$$

Example: Show that $(n+1)^n P_r = {}^{n+1}P_{p+1}$.

Solution: $(n+1)^n P_r = (n+1)\dfrac{n!}{(n-r)!} = \dfrac{(n+1)n!}{(n-r)!}$

$$= \frac{(n+1)!}{[(n+1)-(r+1)]!}\left[\text{writing } n-r \text{ as } [(n+1)-(r+1)]\right]$$

$$= {}^{n+1}P_{r+1} \quad \text{(By definition)}$$

Permutations Under Some Conditions

We will now see examples involving permutations with some extra conditions.

Example:

1. Suppose 7 students are staying in a hall in a hostel and they are allotted 7 beds. Among them, Parvin does not want a bed next to Anju because Anju snores. Then, in how many ways can you allot the beds?

Solution: Let the beds be numbered 1 to 7.

Case 1: Suppose Anju is allotted bed number 1.

Then, Parvin cannot be allotted bed number 2.

So Parvin can be allotted a bed in 5 ways.

After alloting a bed to Parvin, the remaining 5 students can be allotted beds in 5! ways. So, in this case the beds can be allotted in 5 × 5! ways = 600 ways.

Case 2 : Anju is allotted bed number 7.

Then, Parvin cannot be allotted bed number 6.

As in Case 1, the beds can be allotted in 600 ways.

Case 3: Anju is allotted one of the beds numbered 2,3,4,5 or 6.

Parvin cannot be allotted the beds on the right hand side and left hand side of Anju's bed. For example, if Anju is allotted bed number 2, beds numbered 1 or 3 cannot be allotted to Parvin.

Therefore, Parvin can be allotted a bed in 4 ways in all these cases.

After allotting a bed to Parvin, the other 5 can be allotted a bed in 5! ways.

Therefore, in each of these cases, the beds can be allotted in 4 × 5! = 480 ways.

∴ The beds can be allotted in,

(2 × 600 + 5 × 480)ways (1200 + 2400) ways = 3600 ways.

2. In how many ways can an animal trainer arrange 5 lions and 4 tigers in a row so that no two lions are together?

Solution: They have to be arranged in the following way:

L	T	L	T	L	T	L	T	L

The 5 lions should be arranged in the 5 places marked 'L'.

This can be done in 5! ways.

The 4 tigers should be in the 4 places marked 'T'.

This can be done in 4! ways.

Therefore, the lions and the tigers can be arranged in 5! × 4! Ways = 2880 ways.

3. There are 4 books on fairy tales, 5 novels and 3 plays. In how many ways can you arrange these so that books on fairy tales are together, novels are together and plays are together and in the order, books on fairytales, novels and plays.

Solution: There are 4 books on fairy tales and they have to be put together.

They can be arranged in 4! ways.

Similarly, there are 5 novels.

They can be arranged in 5! ways.

And there are 3 plays.

They can be arranged in 3! ways.

So, by the counting principle all of them together can be arranged in 4! × 5! × 3! ways = 17280 ways.

Combinations

Let us consider the example of shirts and trousers. There you have 4 sets of shirts and trousers and you want to take 2 sets with you while going on a trip. In how many ways can you do it?

Let us denote the sets by S_1, S_2, S_3, S_4. Then you can choose two pairs in the following ways:

$$1.\{S_1,S_2\} \quad 2.\{S_1,S_3\} \quad 3.\{S_1,S_4\}$$
$$4.\{S_2,S_3\} \quad 5.\{S_2,S_4\} \quad 6.\{S_3,S_4\}$$

[Observe that $\{S_1, S_2\}$ is the same as $\{S_2, S_1\}$]. So, there are 6 ways of choosing the two sets that you want to take with you. Of course, if you had 10 pairs and you wanted to take 7 pairs, it will be much more difficult to work out the number of pairs in this way.

Now as you may want to know the number of ways of wearing 2 out of 4 sets for two days, say Mon-day and Tuesday, and the order of wearing is also important to you. We know that it can be done in $^4P_2 = 12$ ways. But note that each choice of 2 sets gives us two ways of wearing 2 sets out of 4 sets as shown below:

- $\{S_1, S_2\} \to S_1$ on Monday and S_2 on Tuesday or S_2 on Monday and S_1 on Tuesday.
- $\{S_1, S_3\} \to S_1$ on Monday and S_3 on Tuesday or S_3 on Monday and S_1 on Tuesday.
- $\{S_1, S_4\} \to S_1$ on Monday and S_4 on Tuesday or S_4 on Monday and S_1 on Tuesday.
- $\{S_2, S_3\} \to S_2$ on Monday and S_3 on Tuesday or S_3 on Monday and S_2 on Tuesday.
- $\{S_2, S_4\} \to S_2$ on Monday and S_4 on Tuesday or S_4 on Monday and S_2 on Tuesday.
- $\{S_3, S_4\} \to S_3$ on Monday and S_4 on Tuesday or S_4 on Monday and S_3 on Tuesday.

Thus, there are 12 ways of wearing 2 out of 4 pairs.

This argument holds good in general as we can see from the following theorem.

Theorem: Let $n \geq 1$ be an integer and $r \leq n$. Let us denote the number of ways of choosing r objects out of n objects by ^nC_r. Then,

$$^nC_r = \frac{^nP_r}{r!}$$

Proof : We can choose r objects out of n objects in ^nC_r ways. Each of the r objects chosen can be arranged in r! ways. The number of ways of arranging r objects is r!. Thus, by the counting principle, the number of ways of choosing r objects and arranging the r objects chosen can be done in $^nC_r r!$ ways. But, this is precisely ^nP_r . In other words, we have,

$$^nP_r = r!.\ ^nC_r$$

Dividing both sides by r!, we get the result in the theorem.

Here is an example to help you to familiarise yourself with ^nC_r.

Example:

1. Evaluate each of the following:

$$(a)\ ^5C_2 \quad (b)\ ^5C_3$$

$$(b)\ ^5C_3 \quad (d)\ \frac{^6C_3}{^4C_2}$$

Solution:

$$(a)\ ^5C_2 = \frac{^5P_2}{2!} = \frac{5\cdot4}{1.2} = 10$$

$$(b)\ ^5C_3 = \frac{^5P_3}{3!} = \frac{5\cdot4\cdot3}{1.2.3} = 10$$

$$(c)\ {}^4C_3 + {}^4C_2 = \frac{{}^4P_3}{3!} + \frac{{}^4P_2}{2!} = \frac{4.3.2}{1.2.3} + \frac{1.3}{1.2} = 4 + 6 = 10$$

$$(d)\ {}^6C_3 = \frac{{}^6P_3}{3!} = \frac{6.5.4}{1.2.3} = 20 \text{ and } {}^4C_2 = \frac{4.3}{1.2} = 6$$

$$\therefore \frac{{}^6C_3}{{}^4C_2} = \frac{20}{6} = \frac{10}{3}$$

2. Find the number of subsets of the set {1,2,3,4,5,6,7,8,9,10,11} having 4 elements.

Solution: Here the order of choosing the elements doesn't matter and this is a problem in combinations.

We have to find the number of ways of choosing 4 elements of this set which has 11 elements.

This can be done in,

$${}^{11}C_4 = \frac{11.10.9.8}{1.2.3.4} = 33.\text{ ways.}$$

3. 12 points lie on a circle. How many cyclic quadrilaterals can be drawn by using these points?

Solution: For any set of 4 points we get a cyclic quadrilateral. Number of ways of choosing 4 points out of 12 points is ${}^{12}C_4 = 495$. Therefore, we can draw 495 quadrilaterals.

4. In a box, there are 5 black pens, 3 white pens and 4 red pens. In how many ways can 2 black pens, 2 white pens and 2 red pens can be chosen?

Solution: Number of ways of choosing 2 black pens from 5 black pens.

$$= {}^5C_2 = \frac{{}^5P_2}{2!} = \frac{5.4}{1.2} = 10.$$

Number of ways of choosing 2 white pens from 3 white pens.

$$= {}^3C_2 = \frac{{}^3P_2}{2!} = \frac{3.2}{1.2} = 3.$$

Number of ways of choosing 2 red pens from 4 red pens.

$$= {}^4C_2 = \frac{{}^4P_2}{2!} = \frac{4.3}{1.2} = 6$$

∴ By the Counting Principle, 2 black pens, 2 white pens, and 2 red pens can be chosen in 10 × 3 × 6 = 180 ways.

5. A question paper consists of 10 questions divided into two parts A and B. Each part contains five questions. A candidate is required to attempt six questions in all of which at least 2 should be from part A and at least 2 from part B. In how many ways can the candidate select the questions if he can answer all questions equally well?

Solution: The candidate has to select six questions in all of which at least two should be from Part A and two should be from Part B. He can select questions in any of the following ways:

Part A	Part B
2	4
3	3
4	2

If the candidate follows choice (i), the number of ways in which he can do so is,

$$^5C_2 \times {^5C_4} = 10 \times 5 = 50$$

If the candidate follows choice (ii), the number of ways in which he can do so is,

$$^5C_3 \times {^5C_3} = 10 \times 10 = 100.$$

Similarly, if the candidate follows choice (iii), then the number of ways in which he can do so is,

$$^5C_4 \times {^5C_2} = 50.$$

Therefore, the candidate can select the question in 50 + 100 + 50 = 200 ways.

Some Simple Properties of ^nC_r

We will prove some simple properties of ^nC_r which will make the computations of these coefficients simpler. Let us go back again to Theorem. Using relation we can rewrite the formula for ^nC_r as follows:

$$^nC_r = \frac{n!}{r!(n-r)!}$$

Example:

1. Find the value of $^nC_0$.

Solution: Here r = 0.

Therefore, $^nC_0 = \frac{n!}{0!n!} = \frac{1}{0!} = 1,$

since we have defined 0! = 1.

The formula given in Theorem was used in the previous section. As we will see shortly, the formula given in Equation $^nC_r = \frac{n!}{r!(n-r)!}$ will be useful for proving certain properties of ^nC_r,

$$^nC_r = {^nC_{n-r}}$$

This means just that the number of ways of choosingr objects out of n objects is the same as the number of ways of not choosing (n – r) objects out of n objects. In the example described in the introduction, it just means that the number of ways of selecting 2 sets of dresses is the same as the number of ways of rejecting 4 – 2 = 2 dresses. In Example, this means that the number of ways of choosing subsets with 4 elements is the same as the number of ways of rejecting 8 elements since choosing a particular subset of 4 elements is equivalent to rejecting its complement, which has 8 elements.

Let us now prove this relation using Equation ${}^{n}C_r = \frac{n!}{r!(n-r)!}$. The denominator of the right hand side of this equation is r! (n – r)!. This does not change when we replace r by n – r.

$$(n-r)!.[n-(n-r)]! = (n-r)!.r!$$

The numerator is independent of r. Therefore, replacing r by n – r in Equation ${}^{n}C_r = \frac{n!}{r!(n-r)!}$ we get result.

How is the relation 7.9 useful? Using this formula, we get, for example, ${}^{100}C_{98}$ is the same as ${}^{100}C_2$. The second value is much more easier to calculate than the first one.

2. Evaluate:

$$(a)\,{}^{7}C_5 \quad (c)\,{}^{10}C_9$$
$$(b)\,{}^{11}C_9 \quad (d)\,{}^{12}C_9$$

Solution: (a) We have,

$${}^{7}C_5 = {}^{7}C_{7-5} = {}^{7}C_2 = \frac{7.2}{1.2} = 21$$

(b) Similarly ${}^{10}C_9 = {}^{10}C_{10-9} = {}^{10}C_1 = 10$,

$$(c)\ {}^{11}C_9 = {}^{11}C_{11-9} = {}^{11}C_2 = \frac{11.10}{1.2} = 55$$

$$(d)\ {}^{12}C_{10} = {}^{12}C_{12-10} = {}^{12}C_2 = \frac{12.11}{1.2} = 66$$

There is another relation satisfied by ${}^{n}C_r$ which is also useful. We have the following relation:

$${}^{n-1}C_{r-1} + {}^{n-1}C_r = {}^{n}C_r$$

$${}^{n-1}C_{r-1} + {}^{n-1}C_r = \frac{(n-1)!}{(n-r)!(r-1)!} + \frac{(n-1)!}{(n-r-1)!r!}$$
$$= \frac{(n-1)!}{(n-r)(n-r-1)!(r-1)!} + \frac{(n-1)!}{r(n-r-1)!(r-1)!}$$

$$
\begin{aligned}
{}^{n-1}C_{r-1} + {}^{n-1}C_r &= \frac{(n-1)!}{(n-r)!(r-1)!} + \frac{(n-1)!}{(n-r-1)!r!} \\
&= \frac{(n-1)!}{(n-r)(n-r-1)!(r-1)!} + \frac{(n-1)!}{r(n-r-1)!(r-1)!} \\
&= \frac{(n-1)!}{(n-r-1)!(r-1)!}\left[\frac{1}{n-r} + \frac{1}{r}\right] \\
&= \frac{(n-1)!}{(n-r-1)!(r-1)!}\left[\frac{n}{(n-r)r}\right] \\
&= \frac{n(n-1)!}{(n-r)(n-r-1)!r(r-1)!} \\
&= \frac{n!}{(n-r)!r!} = {}^nC_r
\end{aligned}
$$

3. Evaluate:

(a) ${}^6C_2 + {}^6C_1$ (b) ${}^8C_2 + {}^8C_1$
(c) ${}^5C_3 + {}^5C_2$ (d) ${}^{10}C_2 + {}^{10}C_3$

Solution:

(a) Let us use relation ${}^{n-1}C_{r-1} + {}^{n-1}C_r = {}^nC_r$ with $n = 7, r = 2$. So, ${}^6C_2 + {}^6C_1 = {}^7C_2 = 21$

(b) Here n = 9, r = 2. Therefore, ${}^8C_2 + {}^8C_1 = {}^9C_2 = 36$

(c) Here n = 6, r = 3. Therefore, ${}^5C_3 + {}^5C_2 = {}^6C_3 = 20$

(d) Here n = 11, r = 3. Therefore, ${}^5C_3 + {}^5C_2 = {}^6C_3 = 20$

In this example let us calculate the number of subsets of the set, {1, 2, 3, 4, 5, 6, 7, 8, 9, 10, 11}. We can subdivide them into two kinds, those that contain a particular element, say 1, and those that do not contain 1. The number of subsets of the set having 4 elements and containing 1 is the same as the number of subsets of {2, 3, 4, 5, 6, 7, 8, 9, 10, 11} having 3 elements. There are ${}^{10}C_3$ such subsets.

The number of subsets of the set having 4 elements and not containing 1 is the same as the number of subsets of the set {2,3,4,5,6,7,8,9,10,11,} having 4 elements. This is ${}^{10}C_4$. So, the number of subsets of {1, 2, 3, 4, 5, 6, 7, 8, 9, 10, 11} having four elements is ${}^{10}C_3 + {}^{10}C_4$. But, this is ${}^{11}C_4$ as we have seen in the example. So, ${}^{10}C_3 + {}^{10}C_4 = {}^{11}C_4$. The same argument works for the number of r–element subsets of a set with n elements.

This relation was noticed by the French Mathematician Blaise Pascal and was used in the so called Pascal Triangle.

n = 0						1					
n = 1					1		1				
n = 2				1		2		1			
n = 3			1		3		3		1		
n = 4		1		4		6		4		1	
n = 5	1		5		10		10		5		1

The first row consists of single element ${}^{0}C_0 = 1$. The second row consists of ${}^{1}C_0$ and ${}^{1}C_1$. From the third row onwards, the rule for writing the entries is as follows. Add consecutive elements in the previous row and write the answer between the two entries. After exhausting all possible pairs, put the number 1 at the begining and the end of the row. For example, the third row is got as follows. Second row contains only two elements and they add up to 2. Now, put 1 before and after 2. For the fourth row, we have 1 + 2 = 3, 2 + 1 = 3. Then, we put 1 + 2 = 3, 2 + 1 = 3. Then we put 1 at the beginning and the end. Here, we are able to calculate, for example, ${}^{3}C_1, {}^{3}C_2$., from ${}^{2}C_0, {}^{2}C_1, {}^{2}C_2$ by using the relation 7.10. The important thing is we are able to do it using addition alone.

The numbers ${}^{n}C_r$ occur as coefficents in the binomial expansions, and therefore, they are also called binomial coefficents as we will see in lesson 8. In particular, the property 7.10 will be used in the proof of the binomial theorem.

4. If ${}^{n}C_{10} = {}^{n}C_{12}$ find n,

Solution: Using ${}^{n}C_r = {}^{n}C_{n-r}$ we get,

$n - 10 = 12$

or, $n = 12 + 10 = 22$

5. There are 5 novels and 4 biographies. In how many ways can 4 novels and 2 biographies can be arranged on a shelf?

Solution: 4 novels can be selected out of 5 in ${}^{5}C_4$ ways. 2 biographies can be selected out of 4 in ${}^{4}C_2$ ways.

Number of ways of arranging novels and biographies,

$$= {}^{5}C_2 \times {}^{4}C_2 = 5 \times 6 = 30$$

After selecting any 6 books (4 novels and 2 biographies) in one of the 30 ways, they can be arranged on the shelf in 6! = 720 ways.

By the Counting Principle, the total number of arrangements = 30 × 720 = 21600.

6. From 5 consonants and 4 vowels, how many words can be formed using 3 consonants and 2 vowels?

Solution: From 5 consonants, 3 consonants can be selected in ${}^{5}C_3$ ways.

From 4 vowels, 2 vowels can be selected in ${}^{4}C_2$ ways.

Now with every selection, number of ways of arranging 5 letters is ${}^{5}P_{5}$,

$\therefore$ Total number of words $= {}^{5}C_{3} \times {}^{4}C_{2} \times {}^{5}P_{5}$,

$$= \frac{5\times4}{2\times1} \times \frac{4\times3}{2\times1} \times 5!$$

$$= 10\times6\times5\times4\times3\times2\times1 = 7200$$

Relation between Permutation and Combination

The word permutation means arrangement of the alike or different objects taken some or all at a time. So we can observe the word 'arrangement' used in the definition of permutation. Here the arrangement means selection as well as ordering. That means the order in which the objects are selected have also been taken care of in this case.

Example – The number of 5 digit numbers which can be formed using the digits 0, 1, 2, 3, 4 and 5.

In this example, we just not have to select the 5 digits out of given 6 digits but also have to see the number of possible cases for the different arrangement. So the numbers 34251, 21034, 42351 are all different cases.

The very basic difference in permutation and combination is the order of the objects considered. In combination, the order is not considered at all while for permutation it is must. So the permutation is the ordered arrangement while the combination is the unordered selection.

From the three alphabets A, B and C, the permutation of these 3 letters will be ABC, ACB, BAC, BCA, CBA and CAB. While the combination of 3 letters will be just (A, B, C).

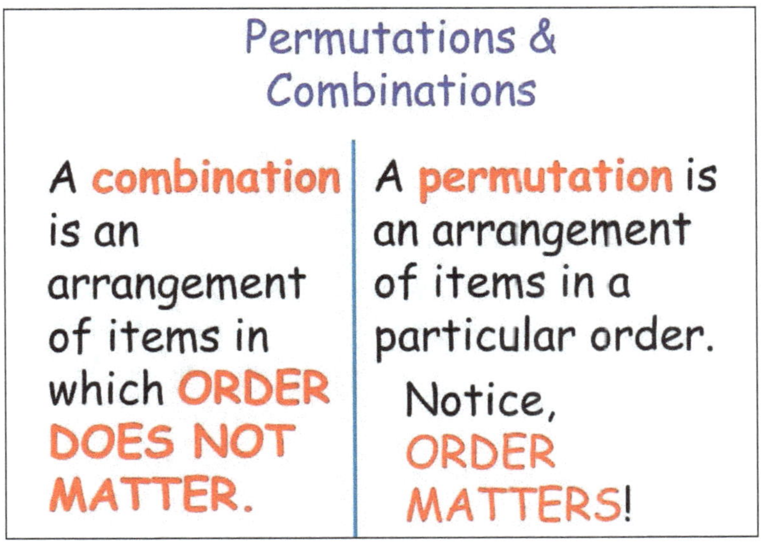

Permutation gives the answer to the number of arrangements while the combination explains the possible number of selections.

Permutation of a single combination can be multiple but the combination of a single permutation is unique (considering all at a time).

Permutation is the combination (or selection) and the arrangement as well. Thus, while calculating the permutation, we first need to choose or selecting the thing before their arrangement. So, Permutation = Selection x Arrangement.

This can also be understood from their mathematical relation. Since we know that,

$$^{n}P_{r} = \frac{n!}{(n-r)!} \text{ and } ^{n}C_{r} = \frac{n!}{(n-r)!r!}$$

Thus, from the above two formulas, this is very clear that,

$$^{n}P_{r} = {}^{n}C_{r}\, r!$$

where, $^{n}C_{r}$ denotes the selection and r! denotes the arrangement of r objects for the r places.

References

- Probability-And-Stats: ucsd.edu, Retrieved 29, June 2020
- Probability: utdallas.edu, Retrieved 09, March 2020
- Cetking-Permutation-Combination-Basic-handout: cetking.com, Retrieved 20, July 2020
- Permutation-and-Combination: eckovation.com, Retrieved 11, January 2020

3

Probability Distributions

The mathematical function which provides the probabilities of the occurrence of different outcomes for a particular trial is known as probability distribution. The topics elaborated in this chapter will help in gaining a better perspective about the major concepts studied in relation to probability distributions such as probability functions of discrete random variables and continuous probability distributions.

Random Variables

Suppose the number of students in a class is 50 that shows whole number and height of students that may lie in some interval because it may be 4.5 or 6.1 or 5.5 inch. A random variable may be associated with either a finite countable number (as well as infinite countable) i.e. discrete value or in finite uncountable interval i.e. continuous values. Random variable is classified as of two types first is discrete and second one is continuous values of variable.

Discrete Random Variable

Discrete data means integer data that mean it does not include decimal values and considered only whole values. Discrete values can be positive or negative. For example: ±1, ±2, ... etc.

"Discrete random variable is a random variable which assume only finite and countable values". List of some examples of discrete random variable:

- Number of phone calls in a day.
- Number of road accidents in a day.
- Number of claims in a day.
- Number of job seekers for a data scientist job.
- Number of grammatical errors in a text.
- Number of customers in a show room.
- Number of defective products in a manufacturer company.
- Number of produced products by a manufacturer company.

Continuous Random Variable

Continuous data means real numbers that include decimal values. Continuous values are like as 2.3, 5.6, 9.8, 8.7.... or any value in any interval (1, 3), (7,9).....etc. These interval may contain infinite values between interval limits.

"Continuous random variable is a random variable which can take any value in a given interval".

List of some examples of continuous random variable:

- Closing price of share market.
- Temperature of metropolitan cities on a particular day.
- Distance travelled by a truck driver.

Probability Distribution Function

Frequency Distribution is used to arrange repetitive data in an appropriate manner. For example: claims reported each day for one week is given as:

0, 1, 2, 1, 3, 4, 1, 0, 2, 5, 4, 3, 7, 6, 6, 5, 2, 4

This data set have repetition of some values. In this data set 0 occurred 2 times, it means frequency of 0 is 2. Frequency denotes the number of data values occur. Another representation of the data set is called frequency distribution.

No of claims reported each day	0	1	2	3	4	5	6	7
Frequency	2	3	3	2	3	2	2	1

This frequency distribution shows the distribution between frequencies shared in the data set values. Probability distribution shows the distribution between probabilities and data values like frequency distribution.

For example: Tossing a coin two times and random variable X denotes the number of heads, Probability distribution of X is given as:

Possible outcomes	HH	HT	TH	TT
X	2	1	1	0
PX	¼	¼	1/4	¼

$$P(X=0)=\frac{1}{4}; \quad P(X=1)=\frac{1}{4}+\frac{1}{4}=\frac{2}{4}=\frac{1}{2}; \quad P(X=2)=\frac{1}{4}.$$

"Probability Distribution is a function which distribute the probabilities of a random variable to each outcome of random variable".

Mathematical Representation

Suppose random variable is X and outcomes as z, then probability distribution function (pdf) is given as:

$$P(X \le x) = P\{z : X(z) \le x\}$$

where, $-\infty < x < \infty$

This is also called Distribution Function and Cumulative Distribution Function.

Probability Mass Function

If X is a discrete random variable and z's are outcomes, then function $p_x(z)$ is given as:

$$p_x(z) = \begin{cases} p(z_i) & \text{if } z = z_i \ \forall\, i = 1, 2, \ldots \\ 0 & \text{otherwise} \end{cases}$$

Function $p_x(z)$ is called probability mass function (PMF) of random variable X.

A function will be called probability mass function, if it follows following conditions:

(i) $p(z_i) \geq 0$ for all i;

(ii) $\sum_{i=1}^{\infty} p(z_i) = 1.$

For example: Rolling a dice which has 6 possible outcome so S={1, 2, 3, 4, 5, 6}. Let X be the random variable defined by:

$$X = \begin{cases} 1 & \text{if even number face of dice occur} \\ 0 & \text{otherwise} \end{cases}$$

$$P(X=1) = \frac{3}{6} = \frac{1}{2};$$

$$P(X=0) = \frac{3}{6} = \frac{1}{2}.$$

One can see that this function is probability mass function, as it satisfies the condition of probability mass function.

(i) $p(z_i) \geq 0$ for all i;

(ii) $\sum_{i=1}^{\infty} p(z_i) = 1.$

Graphical Representation

The graphical representation of probability mass function of the random variable is given in the figure:

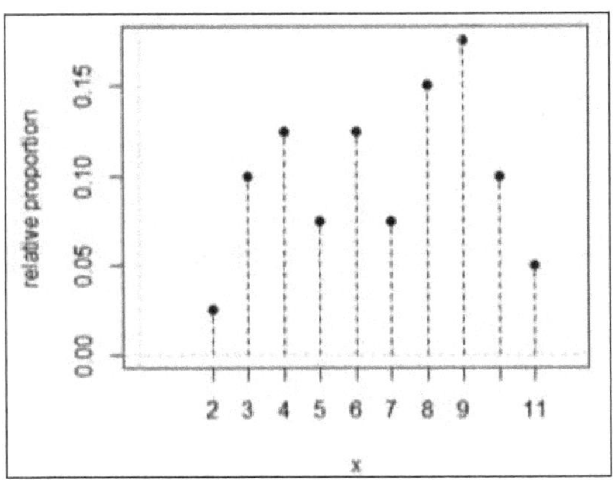

From the figure, we can see that on the x-axis we have discrete point 2, 3,...,11 and on the y-axis the corresponding probabilities are given. So if add up all the probabilities the sum should be equal to one.

Graphical representation of Distribution function for the discrete random variable is presented in the figure.

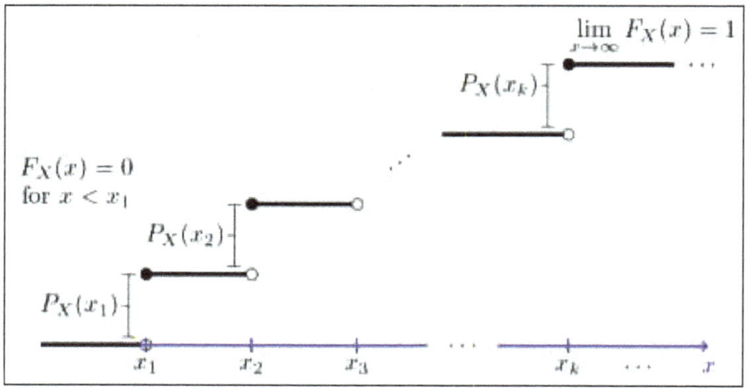

Some question with solutions are given here for better understanding.

Example:

Two dice are rolled. Let X be a random variable which is sum of upturned faces is even.

- Construct probability distribution table.
- Find the distribution function of X for 1 i.e. F(1).

Example:

Two dice are rolled in a random experiment.

Sample space is:

S ={(1,1),(1,2),.....,(1,6),(2,1),(2,2),.....,(2,6),(3,1),(3,2),......,(3,6),(4,1),(4,2),.....,(4,6),

(5,1),(5,2),.......,(5,6),(6,1),(6,2),.....,(6,6)}

Random variable X is given as:

$$X = \begin{cases} 1 & \text{if sum is even} \\ 0 & \text{if sum is odd} \end{cases}$$

Outcomes of random variable X are:

(X = 1) = 18; (X = 0) = 18.

Probability mass function of the random variable is given as:

$$P(X=0) = \frac{18}{36}$$

$$p(X=1) = \frac{18}{36}.$$

Function satisfies the condition of distribution function:

(i) $p(z_i) \geq 0$ for all i;

(ii) $\sum_{i=1}^{\infty} p(z_i) = 1.$

This function is distribution function:

- Probability mass function table is given as:

X	0	1
p(z)	18/36	18/36

- Distribution function is given as:

$$F(1) = P(X \leq 1) = \frac{36}{36} = 1.$$

Example:

A random variable X has a following distribution function:

$$F(z) = \begin{cases} 0 & \text{if } z = 0 \\ \dfrac{18}{36} & \text{if } 0 \leq z < 1 \\ 1 & \text{if } z \geq 1 \end{cases}$$

X	0	1	2	3	4	5	6
p(z)	2	2a	2a	2a*a	a*a	a*a + a	3a*a

- Find value of a.
- Draw chart of Probability Distribution and discrete Distribution function.

Example:

- Since $\sum_{i=1}^{\infty} p(z_i) = 1;$

$$a + 2a + 2a + 2a^2 + a^2 + a^2 + a + 3a^2 = 1;$$
$$7a^2 + 6a - 1 = 0;$$
$$7a^2 + 7a - a - 1 = 0;$$
$$7a(a+1) - (a+1) = 0;$$
$$(a+1)(7a-1) = 0;$$
$$a = -1;\ a = \frac{1}{7}.$$

Since p(z) cannot be negative so,

$$a = \frac{1}{7}.$$

After putting, $a = \frac{1}{7}$ Probability Distribution function is:

X	0	1	2	3	4	5	6
p(z)	0.14	0.28	0.28	0.04	0.04	0.16	0.06

- The distribution function for X is given as:

X	F(z)
0	0.142857
1	0.428571
2	0.714285
3	0.754285
4	0.79510137
5	0.9583666
6	1

Discrete distribution function of X chart is given as:

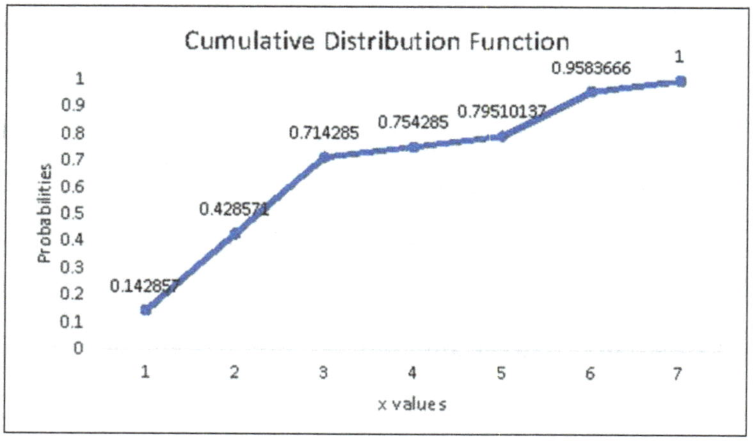

Bar chart of probability function of X variable's is given as:

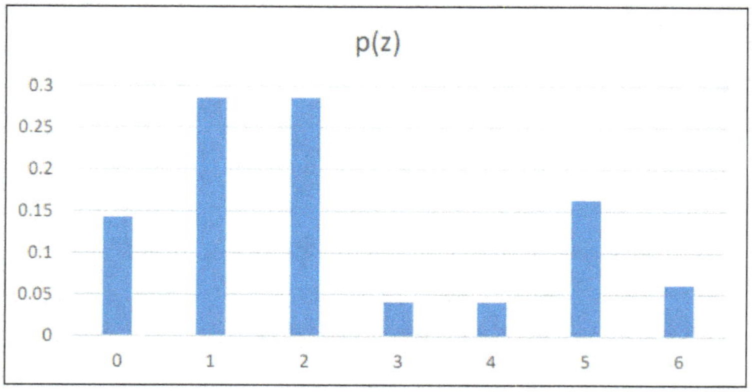

Probability Density Function

If X be a continuous random variable and z outcomes comes from (a, b) interval, area under this interval will be called as probability density function (pdf). Probability Density Function or pdf of X will be shown as $f_x(z)$:

$$P(a \le X \le b) = \int_a^b f(z)dz$$

A function will be PDF in interval, if $(-\infty, \infty)$ function satisfy the following conditions:

(i) $f(z) \ge 0$;

(ii) $\int_{-\infty}^{\infty} f(z)dz = 1$;

(ii) $\int_{-\infty}^{\infty} f(z)dz = 1$;

where E is any event of random experiment.

- Probability value of any constant c is 0 for continuous random variable.

  $P(X = c) = 0, \forall c$

- $P(a \le X \le b) = P(a < X \le b) = P(a \le X < b) = P(a < X < b)$.

Continuous Distribution Function

If X is a continuous random variable with pdf f(z), the function is:

$$F_x(x) = P(X \le x) = \int_{-\infty}^{x} f(z)dz, -\infty < x < \infty.$$

The function $F_x(x)$ is called Distribution function or continuous Distribution function.

Example:

A continuous random variable X has the PDF $f_x(x) = me^{-3z}$; for $z \ge 0$.

Find the value of m and $P(X \le 8)$.

Example:

Since $\int_{-\infty}^{\infty} f(z)dz = 1$;

$$\int_0^{\infty} me^{-3z}\,dz = 1;$$

$$m\int_0^{\infty} e^{-3z}\,dz = 1;$$

$$\left|\frac{me^{-3z}}{-3}\right|_0^{\infty} = 1.$$

After putting the upper and lower limits on the integration:

$$m * \frac{1}{3} = 1;$$
$$m = 3.$$

Now to find the value of $P(Z \leq 8)$:

$$P(Z \leq 8) = \int_{o}^{\infty} 3e^{-3z} dz;$$
$$P(X \leq 8) = \left| \frac{3e^{-3z}}{-3} \right|_{o}^{8}.$$

After putting limits on integration:

$$P(X \leq 8) = \left(-e^{-24} + 1\right);$$
$$P(X \leq 8) = 1 - 3.77513e - 11;$$
$$P(X \leq 8) \approx 1.$$

Mathematical Expectation

Once PDF for the random variable is determined, often one wants to compute expected value or average value of random variable. Expected value for a discrete random variable is a weighted average, where weight is the probabilities to the associated random variable values.

Mathematical expression for the expected value of discrete random variable X with Probability mass function (pmf) f(z):

$$E(X) = \sum_{z} zf(z).$$

Mathematical expression for the expected value of continuous random variable X with pdf f(z):

$$E(X) = \int_{-\infty}^{\infty} zf(z) dz.$$

Expected value for the random variable exist of right hand side of mathematical expression is convergence i.e. finite value of the variable exists.

Properties of Expectation

1. Expected value of function of random variable is same as the expected value of the random variable.

Mathematical expression for the expected value of function of discrete random variable $g(X)$ is given as:

$$E[g(x)] = \sum_{z} g(z) f(z).$$

Mathematical expression for the expected value of function of continuous random variable $g(X)$ is given as:

$$E[g(x)] = \int_{-\infty}^{\infty} g(z) f(z) dz.$$

2. If the expectation of two random variable X and Y exists, expectation of addition of two random variable is equal to addition of expectation of respective two random variable.

Mathematical expression is given as:

$$E(X+Y) = E(X) + E(Y).$$

This property also holds for function of random variable.

3. If the expectation of two random variable X and Y exists, expectation of multiplication of two random variable is equal to multiplication of expectation of respective two random variable.

Mathematical expression is given as:

This property also holds for function of random variable.

$$E(XY) = E(X)\,E(Y).$$

4. Expectation of any constant is equal to the constant. Let us suppose c is any constant value then mathematical expression is given as:

$$E(c) = c.$$

5. Expectation of linear combination of random variable or function of random variable is given as:

$$E(aX + c) = aE(X) + c.$$

where a and c are two constants and X is a random variable.

Probability Functions of Discrete Random Variables

Uniform and Geometric Distribution

To work out the probability that a discrete random variable X takes a particular value x, we need to identify the event (the set of possible outcomes) that corresponds to "X = x".

In general, the function used to describe the probability distribution of a discrete random variable is called its probability function (abbreviated as pf). The probability function of X is the function $p_X : \mathbb{R} \rightarrow [0, 1]$ given by,

$$p_X(x) = \Pr(X = x).$$

In general, the probability function $p_X(x)$ may be specified in a variety of ways. One way is to specify a numerical value for each possible value of X; we have done that for the die-rolling example. In the die-rolling example, the random variable X can take exactly six values, and no others, and we assert that the probability that X takes any one of these values is the same, namely $\frac{1}{6}$.

As is the case generally for functions, the lower-case x here is merely the argument of the function. If we write Pr(X = y) it is essentially the same function, just as $f(x)=2x^2+3x-1$ and $f(y)=2y^2+3y-1$ are the same function. But it helps to associate the corresponding lower-case letter with the random variable we are considering.

Another way to specify the probability function is using a formula.

Less formally, the probability distribution may be represented using a graph, with a spike of height pX (x) at each possible value x of X. If there are too many possible values of X for this visual representation to work, we may choose to omit probabilities that are very close to zero; such values will typically be invisible on a graph anyway.

The following graph shows the probability function for the outcome of rolling a fair die:

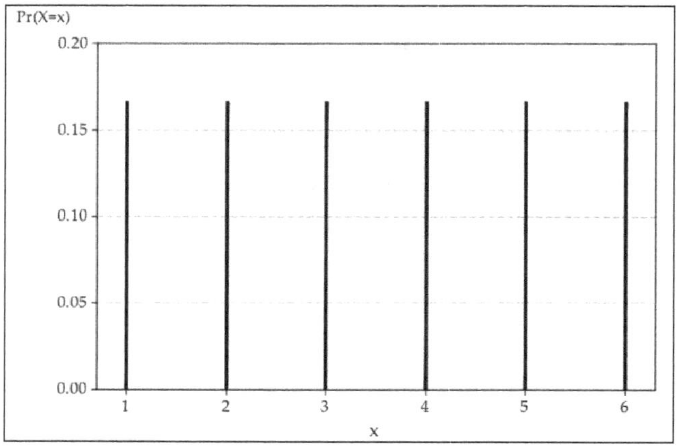

The distribution of X, the number on the uppermost face when a fair die is rolled.

The importance of the notational convention to use capital letters for random variables now becomes apparent. A random variable X has a distribution. Formally, X is a function from the event space to the real line. An observed value of X, such as x = 4, is simply a number. Once we have rolled the die, we have the outcome. If we observed x = 4, this does not have a distribution. It is the observed value, and a number.

Quite often, there is more than one random variable being considered. That is the reason for writing p_X (x) for the probability function of X, to distinguish it from p_Y (y), say, the probability function of Y.

Since the probability function p_X (x) is a probability, it must obey the axioms. Two important properties of a probability function follow from this:

- $p_x(x)=pr(X=x)\geq 0$, for every real number x.
- $\sum p_X(x)=1$, where the sum is taken over all values x for which $p_X(x)>0$.

In fact, these two properties characterise probability functions. If a function f : $\mathbb{R} \to \mathbb{R}$ satisfies these two properties, then we can view f as the probability function of a discrete random variable.

Random variables arising in practical settings of real importance are often much less readily dealt with than those arising in dice rolling. A common approach in such situations, which directly mirrors one of the approaches to probability, is to estimate the probability distribution of the random variable from data.

Example:

Consider the random variable X defined to be the number of languages in which a randomly chosen Australian child attending school can hold an everyday conversation. Suppose we take a random sample of 1000 Australian school children and obtain the following data for the number of languages spoken.

x	1	2	3	4	5	6	Total
Frequency	663	226	66	22	19	4	1000

Then we obtain an estimate of p_X (x), the pf of X, from the relative frequencies.

x	1	2	3	4	5	6
Estimate of p_X (x)	0.663	0.226	0.066	0.022	0.019	0.004

So we estimate that 22.6% of Australian school children can hold an everyday conversation in (exactly) two languages; equivalently, we estimate that Pr (X = 2) = 0.226.

In this case, we cannot be sure that x = 7 (for example) is an impossible value. It may just be so rare that it did not crop up in the sample.

Examples of Discrete Distributions

Discrete Uniform Distribution

Let's consider the random variable defined as the number on the uppermost face when a fair die is rolled. This is an example of a random variable with a discrete uniform distribution. In general, for a positive integer m, let X be a discrete random variable with pf p_X (x) given by,

$$p_X(x) = \Pr(X = x) = \begin{cases} \dfrac{1}{m} & \text{if } x = 1,2,....,m \\ 0 & \text{otherwise.} \end{cases}$$

Then X has a discrete uniform distribution. This is a distribution that arises often in lotteries and games of chance.

In commercial lotteries, such as Powerball, it is a regulatory requirement that each outcome is equally likely. There are 45 possible Powerball numbers (1, 2,..., 45). So if X is the Powerball drawn on a particular occasion, the pf p_X (x) of X is given by,

$$p_X(x) = \Pr(X = x) = \begin{cases} \dfrac{1}{45} & \text{if } x = 1,2,....,45, \\ 0 & \text{otherwise} \end{cases}$$

If this model for the drawing of Powerball numbers is correct, we should expect that, over a large number of draws, the relative frequencies of the 45 possible numbers are all approximately equal to $\frac{1}{45} \approx 0.0222$. The following graph shows the relative frequencies observed in the 853 draws from May 1996 to September 2012.

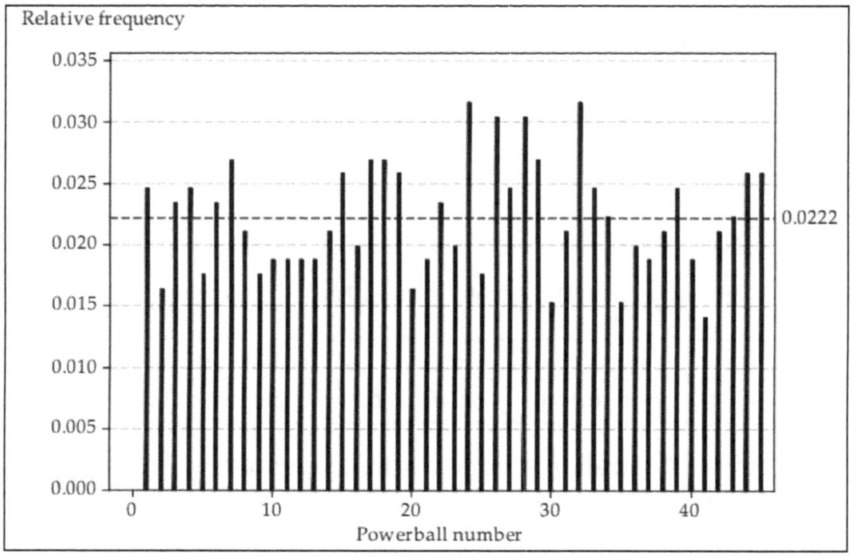

Relative frequencies of Powerball numbers over 853 draws.
The model probability of 0.0222 is shown as a reference line.

Geometric Distribution

The set of possible values for a discrete random variable may be infinite, as the following example shows.

Example:

Fix a real number p with $0 < p < 1$. Now let X be the discrete random variable with probability function $p_X(x)$ given by,

$$p_X(x) = \Pr(X = x) = \begin{cases} p(1-p)^x & \text{if } x = 0,1,2,\ldots, \\ 0 & \text{otherwise.} \end{cases}$$

We will confirm that $p_X(x)$ has the two requisite properties of a probability function. First, we have $p_X(x) \geq 0$, since $p > 0$ and $1 - p > 0$. Second, we have,

$$\begin{aligned} \sum p_X(x) &= p + p(1-p) + p(1-p)^2 + \ldots \\ &= p\left(1 + (1-p) + (1-p)^2 + \ldots\right) \\ &= p\left(\frac{1}{1-(1-p)}\right) \qquad \text{(sum of an infinite geometric series)} \\ &= 1. \end{aligned}$$

Suppose that a sequence of independent 'trials' occur, and at each trial the probability of 'success' equals p. Define X to be the number of trials that occur before the first success is observed. Then X has a geometric distribution with parameter p.

We introduce here a symbol used throughout the modules on probability and statistics. If X has a geometric distribution with parameter p, we write $X \overset{d}{=} G(p)$. The symbol $\overset{d}{=}$ stands for 'has the distribution', meaning the distribution indicated immediately to the right of the symbol.

Note the use of the generic terms 'trial' and 'success'. They are arbitrary, but they carry with them the idea of each observation involving a kind of test (i.e., trial) in which we ask the question: Which one of the two possibilities will be observed, a 'success' or a 'failure'? In this sense, the words 'success' and 'failure' are just labels to keep track of the two possibilities for each trial.

Note that X can take the value 0, if a success is observed at the very first trial. Or it can take the value 1, if a failure is observed at the first trial and then a success at the second trial. And so on. What is the largest value that X can take? There is no upper limit, in theory. As the values of X increase, the probabilities become smaller and smaller.

Another important distribution arises in the context of a sequence of independent trials that each have the same probability of success p. This is the binomial distribution.

Mean of a Discrete Random Variable

If you roll a fair die many times, what will be the average outcome? Imagine rolling it 6000 times. You would expect to roll about 1000 ones, 1000 twos, and so on: about 1000 occurrences of each possible outcome. What would be the average value of the outcomes obtained? Approximately, the average or mean would be:

$$\frac{(1000\times 1)+(1000\times 2)+\cdots+(1000\times 6)}{6000}=\frac{21000}{6000}=3.5$$

This can be thought of as the weighted average of the six possible values 1, 2,..., 6, with weights given by the relative frequencies. Note that 3.5 is not a value that we can actually observe.

By analogy with data and relative frequencies, we can define the mean of a discrete random variable using probabilities from its distribution, as follows:

The mean μ_X of a discrete random variable X with probability function $p_X(x)$ is given by,

$$\mu_X=\sum x p_X(x),$$

where the sum is taken over all values x for which $p_X(x) > 0$.

The mean can be regarded as a measure of 'central location' of a random variable. It is the weighted average of the values that X can take, with weights provided by the probability distribution.

The mean is also sometimes called the expected value or expectation of X and denoted by E(X). These are both somewhat curious terms to use; it is important to understand that they refer to the long-run average. The mean is the value that we expect the long-run average to approach. It is not the value of X that we expect to observe.

Consider a random variable U that has the discrete uniform distribution with possible values 1, 2,...,m. The mean is given by,

$$\begin{aligned}\mu_U &= \sum_{x=1}^{m}\left(x \times \frac{1}{m}\right)\\ &= \frac{1}{m}\sum_{x=1}^{m} x\\ &= \frac{1}{m} \times \frac{m(m+1)}{2}\\ &= \frac{m+1}{2}.\end{aligned}$$

For example, the mean for the roll of a fair die is $\frac{6+1}{2} = 3.5$, as expected.

So in the long run, rolling a single die many times and obtaining the average of all the outcomes, we 'expect' the average to be close to 3.5, and the more rolls we carry out, the closer the average will be.

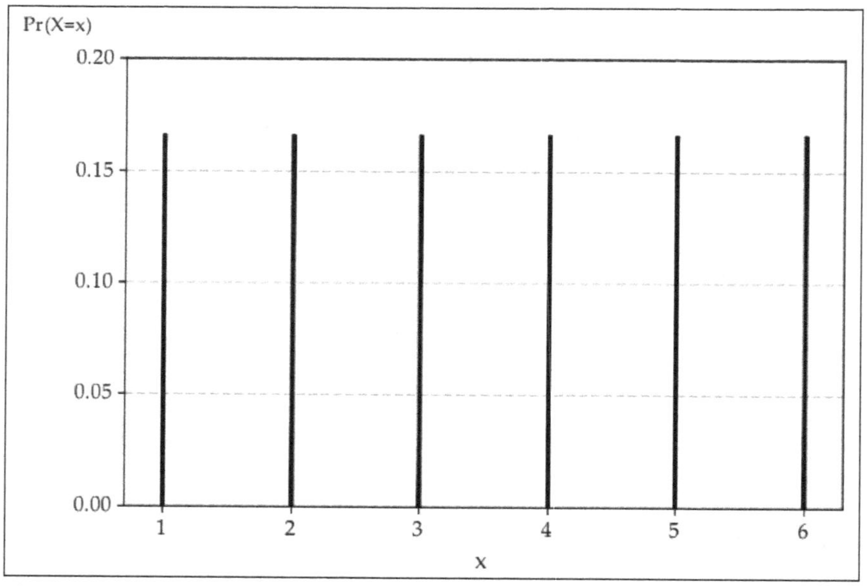

The distribution of X, the number on the uppermost face when a fair die is rolled.

The use of the terms 'expected value' and 'expectation' is the reason for the notation E(X), which also extends to functions of X.

The following graph shows once again the probability function for the outcome of rolling a fair die. This distribution is symmetric, and the mean 3.5 is in the middle of the distribution; in fact, it is on the axis of symmetry.

We can give a general physical interpretation of the mean of a discrete random variable X with pf $p_X(x)$. Suppose we imagine that the x-axis is an infinite see-saw in each direction, and we place weights equal to $p_X(x)$ at each possible value x of X. Then the mean μX is at the point which will make the see-saw balance. In other words, it is at the centre of mass of the system.

If the distribution of a discrete random variable is represented graphically, then you should be able to guess the value of its mean, at least approximately, by using the 'centre of mass' idea.

Suppose that X has a geometric distribution with parameter p, and therefore its probability function is:

$$p_X(x) = p(1-p)^x, \quad x = 0,1,2,....$$

Recall that X is the number of trials before the first success in a sequence of independent trials, each with probability of success p. Do you expect there to be many trials before the first success, on average, or just a few?

A result which we state here without proof is that, for $X \overset{d}{=} G(p)$, we have,

$$\mu_X = \frac{1-p}{p}.$$

If p is large (that is, close to 1), then successes are very likely and the wait before the first success is likely to be short; in this case, μ_X is small. On the other hand, if p is small (close to 0), then failures are very likely and the wait before the first success is likely to be long; in this case, μ_X is large.

We may wish to find the mean of a function of a random variable X, such as X^2 or log X. For a discrete random variable X with pf $p_X(x)$, consider an arbitrary function of X, say Y = g (X). Then the expectation of Y, that is, E(Y) = μ_Y, is obtained as follows:

$$\begin{aligned}\mu_Y &= \sum y P_r(Y=y) && \text{(summing over y for which } Pr(Y = y) > 0)\\ &= \sum g(x)\ Pr(X=x) && \text{(summing over x for which } Pr(X = x) > 0)\\ &= \sum g(x)\ p_X(x).\end{aligned}$$

For the special case of a linear transformation Y = aX +b, we shall see that it follows that $\mu_Y = a\mu_X + b$. This is a very useful result; it says that, for a linear transformation, the mean of the transformed variable equals the transformation of the mean of the original variable. In particular, if Y = aX, then $\mu_Y = a\mu_X$, as you might expect. This applies to changes of units: for example, if the random variable X measures a time interval in days, and we wish to consider the equivalent time in hours, then we can define Y = 24X and we know that $\mu_Y = 24\mu_X$.

For a transformation Y = g (X), it is not true in general that $\mu_Y = g(\mu_X)$. But in the special case of a linear transformation Y = aX + b, where g (x) = ax + b, we have,

$$\begin{aligned}\mu_Y &= \sum g(x) p_X(x)\\ &= \sum (ax+b) p_X(x)\end{aligned}$$

$$= \sum ax\, p_X(x) + \sum b\, p_X(x)$$
$$= a\sum x\, p_X(x) + b\sum p_X(x)$$
$$= a\mu_X + b,$$

Every discrete random variable X with a finite set of possible values has a mean μX. But it is possible to construct examples where the mean does not exist. For example, consider the discrete random variable X with probability function $p_X(x) = \frac{1}{x}$, for x = 2, 4, 8, 16,.... Here $\sum p_X(x) = \sum_{n=1}^{\infty} \frac{1}{2^n} = 1$, as required for a probability function. But $\mu_X = \sum x\, p_X(x) = \sum_{n=1}^{\infty} 1$ does not exist. Such complications are not considered in secondary school mathematics.

Variance of a Discrete Random Variable

We have seen that the mean of a random variable X is a measure of the central location of the distribution of X. If we are summarising features of the distribution of X, it is clear that location is not the only relevant feature. The second most important feature is the spread of the distribution.

If values of X near its mean μ_X are very likely and values further away from μ_X have very small probability, then the distribution of X will be closely concentrated around μ_X. In this case, the spread of the distribution of X is small. On the other hand, if values of X some distance from its mean μ_X are likely, the spread of the distribution of X will be large.

These ideas lead to the most important measure of spread, the variance, and a closely related measure, the standard deviation. Students have met the concepts of variance and standard deviation when summarising data. These were the sample variance and the sample standard deviation. The difference here is that we are referring to properties of the distribution of a random variable.

The variance of a random variable X is defined by,

$$\text{var}(X) = E\left[(X-\mu)^2\right], \text{ where } \mu = E(X).$$

For a discrete random variable X, the variance of X is obtained as follows:

$$\text{var}(X) = \sum (x-\mu)^2 p_X(x),$$

where the sum is taken over all values of x for which $p_X(x) > 0$. So the variance of X is the weighted average of the squared deviations from the mean μ, where the weights are given by the probability function $p_X(x)$ of X.

The standard deviation of X is defined to be the square root of the variance of X. That is,

$$\text{sd}(X) = \sigma_X = \sqrt{\text{var}(X)}.$$

Because of this definition, the variance of X is often denoted by σ_X^2.

In some ways, the standard deviation is the more tangible of the two measures, since it is in the same units as X. For example, if X is a random variable measuring lengths in metres, then the standard deviation is in metres (m), while the variance is in square metres (m^2).

Unlike the mean, there is no simple direct interpretation of the variance or standard deviation. The variance is analogous to the moment of inertia in physics, but that is not necessarily widely understood by students. What is important to understand is that, in relative terms:

- A small standard deviation (or variance) means that the distribution of the random variable is narrowly concentrated around the mean.
- A large standard deviation (or variance) means that the distribution is spread out, with some chance of observing values at some distance from the mean.

Note that the variance cannot be negative, because it is an average of squared quantities. This is appropriate, as a negative spread for a distribution does not make sense. Hence, $var(X) \geq 0$ and $sd(X) \geq 0$ always.

Bernoulli Distribution

Bernoulli distribution is applicable for those random experiment who have only two outcomes like success and failure, head and tail, odd and even, profit and loss etc. Two outcomes are considered as success and failure of the experiment. For example: tossing a coin is a random experiment and getting head is success and getting tail is failure, rolling a die is a random experiment and getting even number is success and getting odd number is failure.

Suppose p denotes the success of the experiment, q denotes the failure of the experiment, X is a random variable is said to have a Bernoulli Distribution with p.m.f. is given as:

$$P(X=z)=f(x)=\begin{cases} p^z q^{1-z}, & \text{for } z=0,1 \\ 0, & \text{otherwise} \end{cases}$$

where, $q = 1 - p$.

Often, random variable X takes value 1 for success and 0 for failure.

Mean

$$E(X)=p;$$
$$E(X^2)=p.$$

Variance

$$V(X)=E(X^2)-(E(X))^2;$$
$$V(X)=p-p^2$$
$$=p(1-p)=pq.$$

Binomial Distribution

If a random experiment is performed repeatedly, each repetition is called as trial. Trial is called Bernoulli trials if number of trials are finite, trials are independent, probability of success is finite. Probability of success p is the occurrence of an event in a Bernoulli trial and q is probability of failure is the non-occurrence of an event in respective Bernoulli trial. Distribution of repeated Bernoulli trials is said to be Binomial Distribution. Binomial distribution was developed by James Bernoulli in year 1700.

Suppose a random variable X is said to follow a Binomial distribution with probability of success is p, probability of failure is q and n independent Bernoulli trials have probability mass function given as:

where z is the outcome of the event.

$$P(X=z)=\begin{cases} {}^{n}_{z}Cp^{z}q^{n-z}, & z=0,1,2,\ldots,n;\ q=1-p \\ 0, & \text{otherwise} \end{cases}$$

Mean of the Binomial Distribution is given as:

$$E(X)=\sum_{z} Z * {}^{n}_{z}Cp^{z}q^{n-z};$$
$$E(X)=np.$$

Variance of the Binomial distribution is given as:

$$V(X)=E(X^{2})-(E(X))^{2}.$$
$$E(X^{2})=\sum_{z} z^{2} * {}^{n}_{z}Cp^{z}q^{n-z};$$
$$E(X^{2})=n(n-1)p^{2}+np;$$

Hence,

$$V(X)=n(n-1)p^{2}+np-n^{2}p^{2};$$
$$V(X)=npq.$$

- Random Variable X is follow Binomial Distribution with parameter n and p is denoted as $X \sim B(n,p)$.
- Variance of Binomial Distribution is always less than mean of Binomial Distribution.
- If X follows Binomial Distribution with parameter n and p, Y = n – X follows Binomial Distribution with parameter n and q.

- If X and Y are two independent random variable and follows Binomial Distribution with parameter n and p, m and p respectively, X + Y also follows Binomial Distribution with parameter m + n and p. This property is said to be additive property of Binomial Distribution.

Mathematical expression is given as:

$$X \sim B(n,p);\ Y \sim B(m,p);\ X+Y \sim B(n+m,p)$$

Poisson Distribution

When event of a random experiment occurs at random point of time and space and our interest lies in the number of happening an event according to our interest, not in non-happening of event. Happening of an event is based on indefinite trials of an experiment. Suppose number of car accidents in particular time period, number of bus passing a crossing per minute during the festive season, number of faulty products in a lump of 100 etc. These type of problems can be solved by using Poisson distribution. Poisson distribution was developed by French mathematician Simeon Denis Poisson in 1837. Poisson distribution is used to count the happening of rare event in a random experiment.

Poisson distribution is a limiting case of Binomial distribution because Poisson distribution has number of trials indefinitely large, probability of success is very small.

Since, $n \to \infty$ and $p \to 0$

Multiplication of n and p must be a constant. Let a be a constant then,

$$np = a;$$

Probability of success is given as:

$$p = \frac{a}{n}$$

Suppose a random variable X is said to have Poisson distribution with parameter a and its pmf is given as:

$$P(X=z) = \begin{cases} \dfrac{e^{-a}a^{z}}{z!}, & z = 0,1,\ldots;\ a > 0 \\ 0, & \text{otherwise} \end{cases}$$

Mean of the Poisson distribution:

$$E(X) = \sum_{z=0}^{\infty} z\frac{e^{-a}a^{z}}{z!};$$

$$E(X) = a.$$

Variance of the Poisson distribution:

$$E(X^2)=\sum_{z=0}^{\infty} z^2 \frac{e^{-a}a^z}{z!};$$

$$V(X)=E(X^2)-(E(X))^2;$$

$$V(X)=a^2+a-a^2;$$

$$V(X)=a.$$

- If X is an random variable and *a* is constant, X follows Poisson distribution with parameter *a* is denoted as:

  $$X\sim P(a).$$

- If X and Y two independent Poisson distributed with parameters a and b respectively then,

  $$X\sim P(a), Y\sim P(b), X+Y\sim P(a+b).$$

  This property of Poisson distribution is known as additive property of Poisson distribution. This property can be extended to n events.

- Poisson distribution is the only distribution whose mean and variance are same.
- Let's say X and Y are two independent Poisson variables, conditional distribution of X given X + Y is Binomial Distribution.
- Poisson distribution is a limiting case of Binomial Distribution.

Continuous Probability Distributions

A probability distribution in which the random variable X can take on any value (is continuous). Because there are infinite values that X could assume, the probability of X taking on any one specific value is zero. Therefore we often speak in ranges of values ($p(X>0) = .50$). The normal distribution is one example of a continuous distribution. The probability that X falls between two values (a and b) equals the integral (area under the curve) from a to b.

Probability Density Function

$$F(x)=P(a\le x\le b)=\int_a^b f(x)dx\ge 0$$

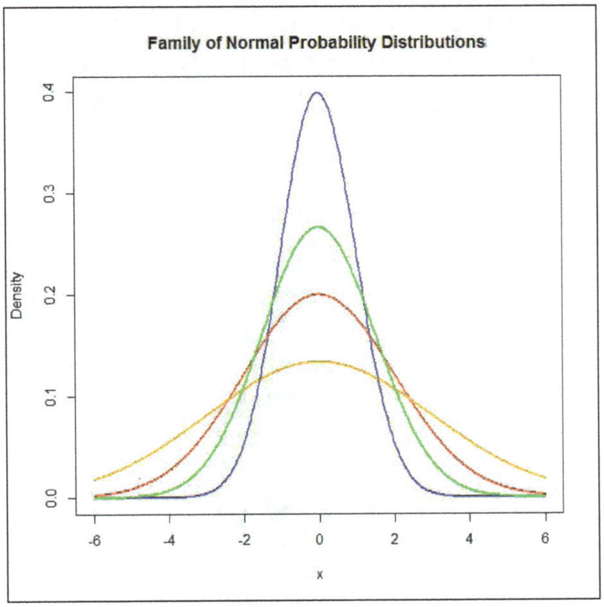

The Normal Probability Distribution

A probability distribution is formed from all possible outcomes of a random process (for a random variable X) and the probability associated with each outcome. Probability distributions may either be discrete (distinct/separate outcomes, such as number of children) or continuous (a continuum of outcomes, such as height). A probability density function is defined such that the likelihood of a value of X between a and b equals the integral (area under the curve) between a and b. This probability is always positive. Further, we know that the area under the curve from negative infinity to positive infinity is one.

The normal probability distribution, one of the fundamental continuous distributions of statistics, is actually a family of distributions (an infinite number of distributions with differing means (μ) and standard deviations (σ). Because the normal distribution is a continuous distribution, we can not calculate exact probability for an outcome, but instead we calculate a probability for a range of outcomes (for example the probability that a random variable X is greater than 10).

The normal distribution is symmetric and centered on the mean (same as the median and mode). While the x-axis ranges from negative infinity to positive infinity, nearly all of the X values fall within +/- three standard deviations of the mean (99.7% of values), while ~68% are within +/-1 standard deviation and ~95% are within +/- two standard deviations. This is often called the three sigma rule or the 68-95-99.7 rule. The normal density function is shown below (this formula won't be on the diagnostic!).

Normal Probability Density Function,

$$F(x) = \frac{1}{\sigma\sqrt{2\pi}} e^{-(x-\mu)^2/2\sigma^2}$$

The standard normal probability function has a mean of zero and a standard deviation of one. Often times the x values of the standard normal distribution are called z-scores. We can calculate probabilities using a normal distribution table (z-table). It is important to note that in these tables, the probabilities are the area to the LEFT of the z-score. If you need to find the area to the right of a z-score (Z greater than some value), you need to subtract the value in the table from one.

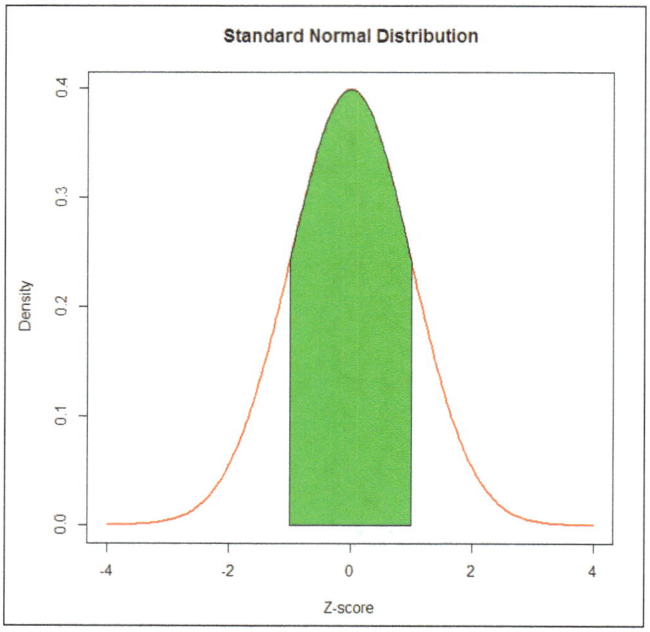

Using this table, we can calculate p(-1<z<1). To do so, first look up the probability that z is less than negative one [p(z)<-1 = 0.1538]. Because the normal distribution is symmetric, we therefore know that the probability that z is greater than one also equals 0.1587 [p(z)>1 = 0.1587]. To calculate the probability that z falls between 1 and -1, we take 1 – 2(0.1587) = 0.6826. The green area in the figure above roughly equals 68% of the area under the curve.

We can convert any and all normal distributions to the standard normal distribution using the equation below. The z-score equals an X minus the population mean (μ) all divided by the standard deviation (σ).

$$Z = \frac{X - \mu}{\sigma}$$

Example of Normal Probability Distribution

We want to determine the probability that a randomly selected blue crab has a weight greater than 1 kg. Based on previous research we assume that the distribution of weights (kg) of adult blue crabs is normally distributed with a population mean (μ) of 0.8 kg and a standard deviation (σ) of 0.3 kg. How do we determine this probability? First, we calculate the z score by replacing X with 1, the mean (μ) with 0.8 and standard deviation (σ) with 0.3. We calculate our z-score to be (1-0.8)/0.3=0.6667. We can then look in our z table to determine the p(z>0.6667) is roughly 1-0.748 (pulled from the chart, somewhere between 0.7454 and 0.7486) = 0.252. Therefore, based on our normality assumption, we conclude that the likelihood that a randomly selected adult blue crab weighs more than one kilogram is roughly 25.2% (the area shaded in blue).

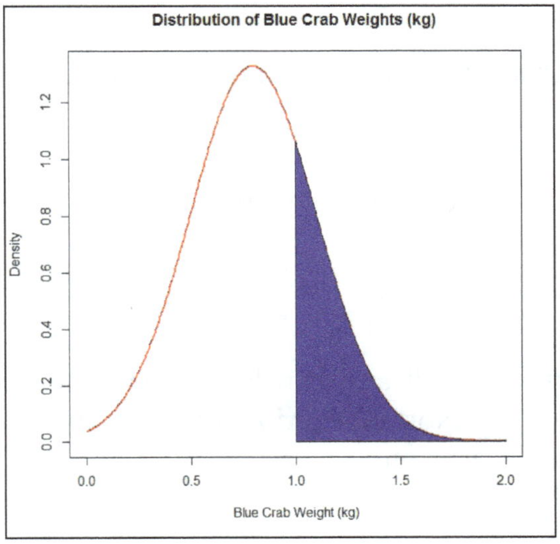

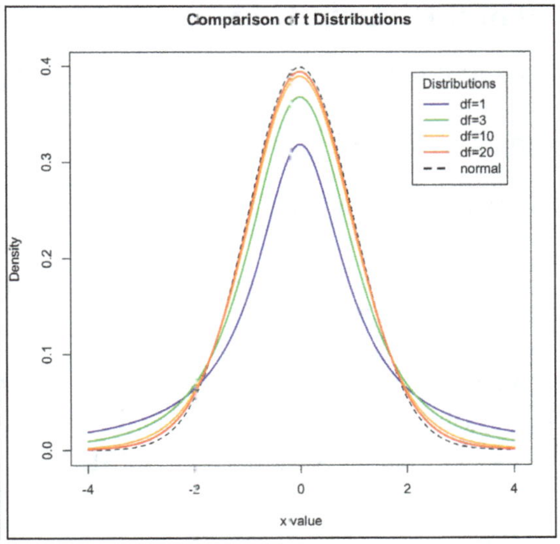

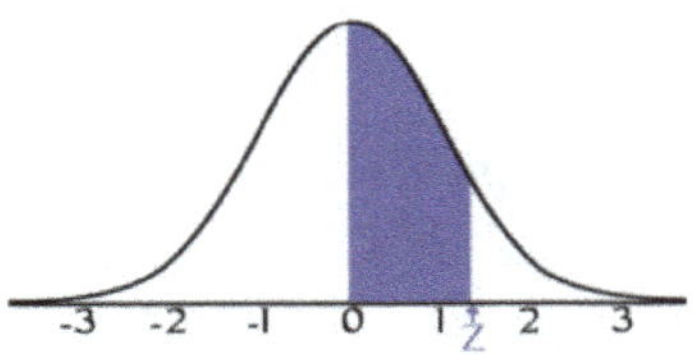

STANDARD NORMAL TABLE (Z)

Entries in the table give the area under the curve between the mean and *z* standard deviations above the mean. For example, for $z = 1.25$ the area under the curve between the mean (0) and *z* is 0.3944.

z	0.00	0.01	0.02	0.03	0.04	0.05	0.06	0.07	0.08	0.09
0.0	0.0000	0.0040	0.0080	0.0120	0.0160	0.0190	0.0239	0.0279	0.0319	0.0359
0.1	0.0398	0.0438	0.0478	0.0517	0.0557	0.0596	0.0636	0.0675	0.0714	0.0753
0.2	0.0793	0.0832	0.0871	0.0910	0.0948	0.0987	0.1026	0.1064	0.1103	0.1141
0.3	0.1179	0.1217	0.1255	0.1293	0.1331	0.1368	0.1406	0.1443	0.1480	0.1517
0.4	0.1554	0.1591	0.1628	0.1664	0.1700	0.1736	0.1772	0.1808	0.1844	0.1879
0.5	0.1915	0.1950	0.1985	0.2019	0.2054	0.2088	0.2123	0.2157	0.2190	0.2224
0.6	0.2257	0.2291	0.2324	0.2357	0.2389	0.2422	0.2454	0.2486	0.2517	0.2549
0.7	0.2580	0.2611	0.2642	0.2673	0.2704	0.2734	0.2764	0.2794	0.2823	0.2852
0.8	0.2881	0.2910	0.2939	0.2969	0.2995	0.3023	0.3051	0.3078	0.3106	0.3133
0.9	0.3159	0.3186	0.3212	0.3238	0.3264	0.3289	0.3315	0.3340	0.3365	0.3389
1.0	0.3413	0.3438	0.3461	0.3485	0.3508	0.3513	0.3554	0.3577	0.3529	0.3621
1.1	0.3643	0.3665	0.3686	0.3708	0.3729	0.3749	0.3770	0.3790	0.3810	0.3830
1.2	0.3849	0.3869	0.3888	0.3907	0.3925	0.3944	0.3962	0.3980	0.3997	0.4015
1.3	0.4032	0.4049	0.4066	0.4082	0.4099	0.4115	0.4131	0.4147	0.4162	0.4177
1.4	0.4192	0.4207	0.4222	0.4236	0.4251	0.4265	0.4279	0.4292	0.4306	0.4319
1.5	0.4332	0.4345	0.4357	0.4370	0.4382	0.4394	0.4406	0.4418	0.4429	0.4441
1.6	0.4452	0.4463	0.4474	0.4484	0.4495	0.4505	0.4515	0.4525	0.4535	0.4545
1.7	0.4554	0.4564	0.4573	0.4582	0.4591	0.4599	0.4608	0.4616	0.4625	0.4633
1.8	0.4641	0.4649	0.4656	0.4664	0.4671	0.4678	0.4686	0.4693	0.4699	0.4706
1.9	0.4713	0.4719	0.4726	0.4732	0.4738	0.4744	0.4750	0.4756	0.4761	0.4767
2.0	0.4772	0.4778	0.4783	0.4788	0.4793	0.4798	0.4803	0.4808	0.4812	0.4817
2.1	0.4821	0.4826	0.4830	0.4834	0.4838	0.4842	0.4846	0.4850	0.4854	0.4857
2.2	0.4861	0.4864	0.4868	0.4871	0.4875	0.4878	0.4881	0.4884	0.4887	0.4890
2.3	0.4893	0.4896	0.4898	0.4901	0.4904	0.4906	0.4909	0.4911	0.4913	0.4916
2.4	0.4918	0.4920	0.4922	0.4925	0.4927	0.4929	0.4931	0.4932	0.4934	0.4936
2.5	0.4938	0.4940	0.4941	0.4943	0.4945	0.4946	0.4948	0.4949	0.4951	0.4952
2.6	0.4953	0.4955	0.4956	0.4957	0.4959	0.4960	0.4961	0.4962	0.4963	0.4964
2.7	0.4965	0.4966	0.4967	0.4968	0.4969	0.4970	0.4971	0.4972	0.4973	0.4974
2.8	0.4974	0.4975	0.4976	0.4977	0.4977	0.4978	0.4979	0.4979	0.4980	0.4981
2.9	0.4981	0.4982	0.4982	0.4983	0.4984	0.4984	0.4985	0.4985	0.4986	0.4986
3.0	0.4987	0.4987	0.4987	0.4988	0.4988	0.4989	0.4989	0.4989	0.4990	0.4990
3.1	0.4990	0.4991	0.4991	0.4991	0.4992	0.4992	0.4992	0.4992	0.4993	0.4993
3.2	0.4993	0.4993	0.4994	0.4994	0.4994	0.4994	0.4994	0.4995	0.4995	0.4995
3.3	0.4995	0.4995	0.4995	0.4996	0.4996	0.4996	0.4996	0.4996	0.4996	0.4997
3.4	0.4997	0.4997	0.4997	0.4997	0.4997	0.4997	0.4997	0.4997	0.4997	0.4998

Log-Normal Distribution

A random variable Y is said to have Log-Normal Distribution if log Y is normally distributed.

Let, $U = \log Y \sim N(\mu, \sigma^2)$, for $y > 0$

The PDF of the Log-Normal distribution of the random variable Y is given as:

$$f_Y(u) = \frac{1}{u\sigma\sqrt{2\pi}} \exp\left\{-(\log u - u)^2 / 2\sigma^2\right\},\ u > 0$$

The volume of gas in a petroleum reserve, amounts of rainfall, size distribution of rainfall droplets, milk production by cows all can be modeled with a Log-Normal distribution. Log-Normal distribution is also known as Galton and Cobb-Douglas distribution.

Properties

- Mean of the Log-Normal distribution:

$$E(u) = e^{\mu + \frac{1}{2}\sigma^2}$$

- Variance of the Log-Normal distribution:

$$V(u) = \left(e^{2\mu+\sigma^2}\right)\left(e^{\sigma^2} - 1\right)$$

Exercise: If Y is log normally distributed with parameters mean 5 and variance 8. What is the value of the probability $P(400 \le Y \le 600)$?

Solution:

$$P(400 \le Y \le 600) = P(Y \le 600) - P(Y \le 400);$$

$$P(Y \le 600) = \phi\left(\frac{\log(600) - 5}{8}\right) = \phi(-0.40) = 0.65542;$$

$$P(Y \le 400) = \phi\left(\frac{\log(400) - 5}{8}\right) = \phi(-0.299) = 0.61791;$$

$$P(400 \le Y \le 600) = P(Y \le 600) - P(Y \le 400)$$
$$= 0.65542 - 0.61791 = 0.03751.$$

Hence the probability value is 0.03751.

Rectangular Distribution

Rectangular Distribution or Uniform Distribution is applicable for both discrete and continuous random variable. In this distribution, all the values that random variable can take in a given interval have an equal probability of occurrence. Discrete uniform distribution is a constant quantity who's PMF depends on the number of integers observations. The probability density function of the uniform distribution is also a constant quantity which depends on the length of given interval. The main application of the uniform distribution that is random number generation as well as p-value. P-value is the value that is used to make conclusion about the acceptance and rejection of the null hypothesis. In case of generating random number from any continuous distribution, first generate the random numbers from the uniform distribution that has interval. Then equate the CDF of that particular distribution with the generated number from the uniform distribution and then find the required random number.

Suppose a random variable Y is said to have rectangular distribution over the given interval (m, n) i.e.

$$Y \sim U(m,n) \text{ or } Y \sim R(m,n).$$

PDF of the rectangular distribution is given as:

$$f_y(Y) = \frac{1}{n-m} \text{ if } m<y<n$$

Cumulative Distribution Function (CDF) $F_Y(y)$ is given as:

$$F_Y(y) = \int_{-\infty}^{y} f_Y(t)\,dt$$

$$F_Y(y) = \frac{y-m}{n-m}, \quad m<y<n$$

$$F_Y(y) = 1, \quad y \geq n$$

Properties of the Rectangular Distribution

- Mean of the rectangular distribution is given as:

$$E(Y) = \frac{m+n}{2}$$

- Variance of the rectangular distribution is given as:

$$V(y) = \frac{(n-m)^2}{12}$$

- Moment generating function is given as:

$$M_y(t) = \frac{\exp(nt)-\exp(mt)}{t(n-m)}$$

- If Y has a uniform distribution in [0, 1], distribution of −2 log Y follows exponential distribution with parameter value $\frac{1}{2}$.

Exercise: If $Y \sim U(8,20)$, what is the value of $P(9 < Y < 18.)$ Also find $E(Y)$, $V(Y)$.

Solution: PDF of the uniform distribution is given as:

$$f_y(Y) = \frac{1}{n-m} \text{if } m < y < n$$

$$f_y(Y) = \frac{1}{20-8} = \frac{1}{12} \text{ if } 8 < y < 20$$

Now, calculate the probability,

$$P(9 < Y < 18) = \int_9^{18} \frac{1}{12} dy$$

$$= \frac{1}{12}[y]_9^{18} = \frac{1}{12}(18-9) = \frac{9}{12} = \frac{3}{4}.$$

Mean of uniformly distributed random variable Y is given as:

$$E(y) = \frac{m+n}{2};$$

$$E(Y) = \frac{20+8}{2} = \frac{28}{2} = 14.$$

Variance of the uniformly distributed random variable Y is given as:

$$V(y) = \frac{(n-m)^2}{12};$$

$$V(y) = \frac{(20-8)^2}{12};$$

$$V(y) = 12.$$

Exercise: The average weight gained by a person in winter is uniformly distributed from 0 to 50lbs. Find the probability a person will gain weight between 17 and 30 lbs. in winter.

Solution: First find the range of the variable (y axis) on the axis. We know that area under a probability distribution is always 1. Hence the random variable follows uniform distribution with constant probability of $\frac{1}{50}$.

Now find the width of the area in which we have to find the probability that is between 17 and 30 is 30-17= 13.

Hence the probability of a person gain weight in winter between 17 and 30 is given as:

Probability value is $13 * \frac{1}{50} = 0.26$.

Beta Distribution of Type I

Beta continuous variable having distribution of type I represents a family of probabilities and outcomes for proportions. For example: How likely is that current Prime Minister of India will win the next prime minister election? One of us few think that he will win, one of us think he will not think and some of them think he will win or not. To solve this type of problem's solution beta distribution will be used. Beta distribution of Type I is also known as basic beta distribution or beta distribution of first kind. There are many applications of the basic beta distribution like testing of Bayesian hypothesis, the rule of succession, task duration modeling etc.

A random variable Y is said to have a beta distribution of first kind with parameter α and β i.e. $Y \sim B_1(\alpha, B)$.

The PDF of the beta distribution of first kind is given as:

$$f(y) = \frac{1}{B(\alpha,\beta)} y^{\alpha-1}(1-y)^{\beta-1}, 0 < y < 1, \alpha, \beta > 0.$$

Properties of the Beta Distribution of First Kind

- Mean of the distribution is given as:

$$E(Y) = \frac{\alpha}{\alpha+\beta}.$$

- Variance of the distribution is given as:

$$V(y) = \frac{\alpha\beta}{(\alpha+\beta)^2(\alpha+\beta+1)}.$$

- Harmonic mean of the distribution is given as:

$$H = \frac{\alpha-1}{\alpha+\beta-1}.$$

- If $\alpha = \beta = 1$, beta distribution of first kind follows uniform distribution with [0, 1].

- If $Y \sim B_1(\alpha, \beta), (1-Y) \sim B_1(B, \alpha)$.

- Suppose Y follows beta distribution with parameter α and β. X is another random variable such that $X | Y \sim B(n, y)$ then $Y | (X = x) \sim B(\alpha + x, \beta + n - x)$.

Exercise: Suppose Y is a random variable with PDF $f(y) = ky^3(1-y)^4, 0 < y < 1$, where k is a constant. Find the value of k.

Solution:

The PDF of the beta distribution is given as:

$$f(y)=\frac{1}{B(\alpha,\beta)}y^{\alpha-1}(1-y)^{\beta-1}, 0<y<1,\ \alpha,\beta>0$$

Given PDF is:

$$f(y)=ky^3(1-y)^4, 0<y<1$$

By comparison of these two PDF, we can see that $\alpha = 4$, $\beta = 5$.

$$k=\frac{1}{B(\alpha,\beta)};$$

$$k=\frac{\Gamma 9}{\Gamma 4\,\Gamma 5}=\frac{40320}{6*24}=280.$$

Beta Distribution of Second Kind or Beta Distribution of Type II

Beta distribution of second kind is also known as beta prime distribution and inverted beta distribution. Beta distribution of second kind is the odds distribution associated with a beta distributed random variable. Same as the basic beta distribution, the beta distribution of second kind is also used to model random probabilities and proportions.

A random variable Y is said to have beta distribution of second kind with parameter α and β i.e. $Y \quad B\ (\alpha,\beta)$.

PDF of the beta distribution of second kind is given as:

$$f(y)=\frac{1}{B(\alpha,\beta)}\frac{y^{\alpha-1}}{(1+y)^{\alpha+\beta}},\ 0<y<\infty,\ \alpha,\ \beta>0.$$

Properties of the Beta Distribution of Second Kind

- Mean of the distribution is given as:

$$E(Y)=\frac{\alpha}{\beta-1}.$$

- Variance of the distribution is given as:

$$V(y)=\frac{\alpha(\alpha-\beta-1)}{(\beta-1)^2(\beta-2)}.$$

- Harmonic mean of the distribution is given as:

$$H=\frac{\alpha-1}{\beta}.$$

- If $Y \sim B_2(\alpha, \beta)$,

$$1+Y=\frac{1}{X}\Rightarrow X=\frac{1}{1+Y}\sim B_1(\beta, \alpha).$$

Exponential Distribution

A continuous random variable X is said to have an Exponential(λ) distribution if it has probability density function:

$$f_X(x\,|\,\lambda)=\begin{cases}\lambda e^{-\lambda x} & \text{for } x>0\\ 0 & \text{for } x\le 0\end{cases},$$

where $\lambda > 0$ is called the rate of the distribution.

In the study of continuous-time stochastic processes, the exponential distribution is usually used to model the time until something happens in the process. The mean of the Exponential(λ) distribution is calculated using integration by parts as:

$$\begin{aligned}E[X]&=\int_0^{\infty} x\lambda e^{-\lambda x}dx\\ &=\lambda\left[\left.\frac{-xe^{-\lambda x}}{\lambda}\right|_0^{\infty}+\frac{1}{\lambda}\int_0^{\infty} e^{-\lambda x}dx\right]\\ &=\lambda\left[0+\frac{1}{\lambda}\left.\frac{-e^{-\lambda x}}{\lambda}\right|_0^{\infty}\right]\\ &=\lambda\frac{1}{\lambda^2}=\frac{1}{\lambda}.\end{aligned}$$

So one can see that as λ gets larger, the thing in the process we're waiting for to happen tends to happen more quickly, hence we think of λ as a rate.

As an exercise, you may wish to verify that by applying integration by parts twice, the second moment of the Exponential(λ) distribution is given by,

$$E\left[X^2\right]=\int_0^{\infty} x^2\lambda e^{-\lambda x}=\ldots=\frac{2}{\lambda^2}.$$

From the first and second moments we can compute the variance as:

$$Var(X)=E\left[X^2\right]-E[X]^2=\frac{2}{\lambda^2}-\frac{1}{\lambda^2}=\frac{1}{\lambda^2}.$$

The Memoryless Property

The following plot illustrates a key property of the exponential distribution. The graph after the point s is an exact copy of the original function. The important consequence of this is that the distribution of X conditioned on {X > s} is again exponential.

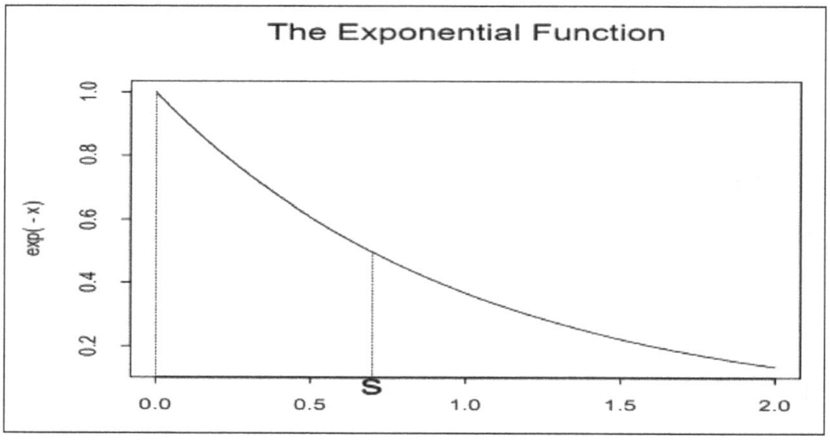

The Exponential Function e^{-x}.

To see how this works, imagine that at time 0 we start an alarm clock which will ring after a time X that is exponentially distributed with rate λ. Let us call X the lifetime of the clock. For any t > 0, we have that:

$$P(X>t)=\int_t^{\infty}\lambda e^{-\lambda x}dx=\lambda\left.\frac{-e^{-\lambda x}}{\lambda}\right|_t^{\infty}=e^{-\lambda t}.$$

Now we go away and come back at time s to discover that the alarm has not yet gone off. That is, we have observed the event {X > s}. If we let Y denote the remaining lifetime of the clock given that {X > s}, then,

$$\begin{aligned}P(Y>t\,|\,X>s)&=P(X>s+t\,|\,X>s)\\&=\frac{P(X>s+t\,|,X>s)}{P(X>s)}\\&=\frac{P(X>s+t)}{P(X>s)}\\&=\frac{e^{-\lambda(s+t)}}{e^{-\lambda s}}\\&=e^{-\lambda t}.\end{aligned}$$

But this implies that the remaining lifetime after we observe the alarm has not yet gone off at time s has the same distribution as the original lifetime X. The really important thing to note, though, is that this implies that the distribution of the remaining lifetime does not depend on s. In fact, if you try setting X to have any other continuous distribution, then ask what would be the distribution of the remaining lifetime after you observe {X > s}, the distribution will depend on s.

This property is called the memoryless property of the exponential distribution because I don't need to remember when I started the clock. If the distribution of the lifetime X is Exponential(λ), then if I come back to the clock at any time and observe that the clock has not yet gone off, regardless of when the clock started I can assert that the distribution of the time till it goes off, starting at the time I start observing it again, is Exponential(λ). Put another way, given that the clock has currently not yet gone off, I can forget the past and still know the distribution of the time from my current time to the time the alarm will go off. The resemblance of this property to the Markov property should not be lost on you.

It is a rather amazing, and perhaps unfortunate, fact that the exponential distribution is the only one for which this works. The memoryless property is like enabling technology for the construction of continuoustime Markov chains. But the exponential distribution is even more special than just the memoryless property because it has a second enabling type of property.

Another Important Property of the Exponential

Let $X_1,...,X_n$ be independent random variables, with X_i having an Exponential (λ_i) distribution. Then the distribution of min $(X_1,....,X_n)$ is Exponential $(\lambda_1+...+\lambda_n)$, and the probability that the minimum is X_i is $\lambda_i/(\lambda_1+...+\lambda_n)$.

Proof:

$$\begin{aligned} P(\min(X_1,...,X_n)>t) &= P(X_1>t,....,X_n>t) \\ &= p(X_1>t)...P(X_n>t) \\ &= e^{-\lambda_1 t}...e^{-\lambda_n t} \\ &= e^{-(\lambda_1+...+\lambda_n)t}. \end{aligned}$$

The preceding shows that the CDF of min $(X_1,....,X_n)$ is that of an Exponential $(\lambda_1+...+\lambda_n)$ distribution. The probability that X_i is the minimum can be obtained by conditioning:

$$P(X_i \text{ is the minimum})$$

$$\begin{aligned} &= P(X_i < X_j \text{ for } j \neq i) \\ &= \int_0^\infty P(X_i < X_j \text{ for } j \neq i \mid X_i = t)\lambda_i e^{-\lambda_i t}dt \\ &= \int_0^\infty P(t < X_j \text{ for } j \neq i)\lambda_i e^{-\lambda_i t}dt \\ &= \int_0^\infty \lambda_i e^{-\lambda_i t}\prod_{j\neq i} P(X_j > t)dt \\ &= \int_0^\infty \lambda_i e^{-\lambda_i t}\prod_{j\neq i} e^{-\lambda_j t}dt \end{aligned}$$

$$= \lambda_i \int_0^\infty e - (\lambda_1 + \ldots + \lambda_n)^t \, dt$$

$$= \lambda_i \left. \frac{-e - (\lambda_1 + \ldots + \lambda_n)}{\lambda_1 + \ldots + \lambda_n} \right|_0^\infty$$

$$= \frac{\lambda_i}{\lambda_1 + \ldots + \lambda_n},$$

as required.

To see how this works together with the the memoryless property, consider the following examples.

Example: Consider a two-server system in which a customer is served first by server 1, then by server 2, and then departs. The service times at server i are exponential random variables with rates μ_i, i = 1, 2. When you arrive, you find server 1 free and two customers at server 2 — customer A in service and customer B waiting in line.

- Find P_A, the probability that A is still in service when you move over to server 2.
- Find P_B, the probability that B is still in the system when you move over to 2.
- Find E[T], where T is the time that you spend in the system.

Solution:

- A will still be in service when you move to server 2 if your service at server 1 ends before A's remaining service at server 2 ends. Now A is currently in service at server 2 when you arrive, but because of memorylessness, A's remaining service is Exponential (μ_2), and you start service at server 1 that is Exponential (μ_1). Therefore, PA is the probability that an Exponential (μ_1) random variable is less than an Exponential (μ_2) random variable, which is:

$$P_A = \frac{\mu_1}{\mu_1 + \mu_2}.$$

- B will still be in the system when you move over to server 2 if your service time is less than the sum of A's remaining service time and B's service time. Let us condition on the first thing to happen, either A finishes service or you finish service:

P(B in system) = P(B in system|A finishes before you) $\frac{\mu_2}{\mu_1 + \mu_2}$ + P(B in system|you finish before A) $\frac{\mu_1}{\mu_1 + \mu_2}$.

Now P(B in system|you finish before A) = 1 since B will still be waiting in line when you move to server 2. On the other hand, if the first thing to happen is that A finishes service, then at that point, by memorylessness, your remaining service at server 1 is Exponential (μ_1), and B will still be in the system if your remaining service at server 1 is less than B's service at server 2, and the probability of this is $\mu_1/(\mu_1 + \mu_2)$. That is:

$$P(\text{B in system}|\text{A finishes before you}) = \frac{\mu_1}{\mu_1 + \mu_2}.$$

Therefore,

$$P(\text{B in system}) = \frac{\mu_1\mu_2}{(\mu_1 + \mu_2)^2} + \frac{\mu_1}{\mu_1 + \mu_2}.$$

- To compute the expected time you are in the system, we first divide up your time in the system into,

  $T = T_1 + R,$

Where T_1 is the time until the first thing that happens, and R is the rest of the time. The time until the first thing happens is Exponential ($\mu_1 + \mu_2$), so that,

$$E[T_1] = \frac{1}{\mu_1 + \mu_2}.$$

To compute E[R], we condition on what was the first thing to happen, either A finished service at server 2 or you finished service at server 1. If the first thing to happen was that you finished service at server 1, which occurs with probability $\mu_1/(\mu_1 + \mu_2)$, then at that point you moved to server 2, and your remaining time in the system is the remaining time of A at server 2, the service time of B at server 2, and your service time at server 2. A's remaining time at server 2 is again Exponential (μ_2) by memorylessness, and so your expected remaining time in service will be $3/\mu_2$. That is,

E[R|first thing to happen is you finish service at server 1] = $\frac{3}{\mu_2}$, and so $E[R] = \frac{3}{\mu_2}\frac{\mu_1}{\mu_1 + \mu_2} + E[R \mid \text{first thing is A finishes}]\frac{\mu_2}{\mu_1 + \mu_2}$.

Now if the first thing to happen is that A finishes service at server 2, we can again compute your expected remaining time in the system as the expected time until the next thing to happen (either you or B finishes service) plus the expected remaining time after that. To compute the latter we can again condition on what was that next thing to happen. We will obtain,

$$E[R|\text{first thing is A finishes}] = \frac{1}{\mu_1 + \mu_2} + \frac{2}{\mu_2}\frac{\mu_1}{\mu_1 + \mu_2} + \left(\frac{1}{\mu_1} + \frac{1}{\mu_2}\right)\frac{\mu_2}{\mu_1 + \mu_2}$$

Plugging everything back gives E[T].

As an exercise you should consider how you might do the preceding problem assuming a different service time distribution, such as a Uniform distribution on [0, 1] or a deterministic service time such as 1 time unit.

Gamma Distribution

A gamma distribution is a general type of statistical distribution that is related to the beta distribution and arises naturally in processes for which the waiting times between Poisson distributed events are relevant. Gamma distributions have two free parameters, labeled α and θ.

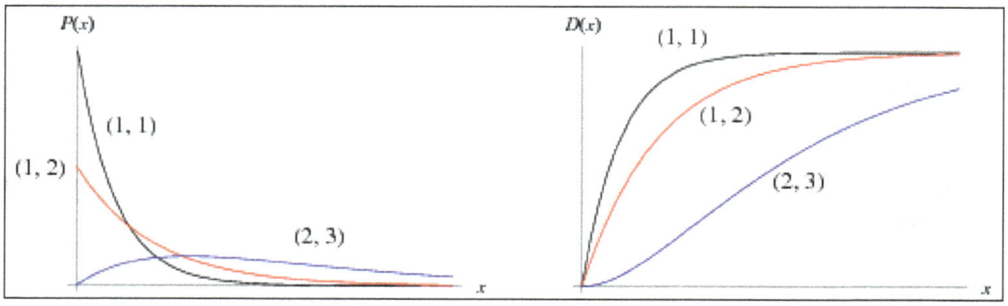

Consider the distribution function D(x) of waiting times until the hth Poisson event given a Poisson distribution with a rate of change λ,

$$
\begin{aligned}
D(x) &= P(X \le x) \\
&= 1 - P(X > x) \\
&= 1\sum_{k=0}^{h-1} \frac{(\lambda x)^k e^{-\lambda x}}{k!} \\
&= 1 - e^{-\lambda x} \sum_{k=0}^{h-1} \frac{(\lambda x)^k}{k!} \\
&= 1 - \frac{\Gamma(h, x\lambda)}{\Gamma(h)}
\end{aligned}
$$

For $x \in [0, \infty)$, where $\Gamma(x)$ is a complete gamma function, and $\Gamma(a, x)$ an incomplete gamma function. With h an integer, this distribution is a special case known as the Erlang distribution.

The corresponding probability function $P(x)$ of waiting times until the hth Poisson event is then obtained by differentiating $D(x)$,

$$
\begin{aligned}
P(x) &= D'(x) \\
&= \lambda e^{-\lambda x} \sum_{k=0}^{h-1} \frac{(\lambda x)^k}{k!} - e^{-\lambda x} \sum_{k=0}^{h-1} \frac{k(\lambda x)^{k-1}\lambda}{k!} \\
&= \lambda e^{-\lambda x} + \lambda e^{-\lambda x} \sum_{k=1}^{h-1} \frac{(\lambda x)}{k!} - e^{-\lambda x} \sum_{k=1}^{h-1} \frac{k(\lambda x)^{k-1}\lambda}{k!} \\
&= \lambda e^{-\lambda x} - \lambda e^{-\lambda x} \sum_{k=1}^{h-1} \left[\frac{k(\lambda x)^{k-1}}{k!} - \frac{(\lambda x)^k}{k!} \right] \\
&= \lambda e^{-\lambda x} \left\{ 1 - \sum_{k=1}^{h-1} \left[\frac{(\lambda x)^{h-1}}{(h-1)!} \right] \right\}
\end{aligned}
$$

$$= \frac{\lambda(\lambda x)^{h-1}}{(h-1)!} e^{-\lambda x}.$$

Now let $\alpha \equiv h$ (not necessarily an integer) and define $\theta \equiv 1/\lambda$ to be the time between changes. Then the above equation can be written:

$$P(x) = \frac{x^{\alpha-1} e^{-x/\theta}}{\Gamma(\alpha)\theta^{\alpha}}$$

for $x \in [0,\infty)$. This is the probability function for the gamma distribution, and the corresponding distribution function is:

$$D(x) = P\left(\alpha, \frac{x}{\theta}\right),$$

Where, $P(a,z)$ is a regularized gamma function.

It is implemented in the Wolfram Language as the function GammaDistribution[alpha, theta].

The characteristic function describing this distribution is:

$$\begin{aligned}\phi(t) &= \mathcal{F}_x\left\{\frac{x^{-x/\theta}\ x^{\alpha-1}}{\Gamma(\alpha)\theta^{\alpha}}\right\}\left[\frac{1}{2}(1+\operatorname{sgn} x)\right](t)\\ &= (1-it\theta)^{-\alpha},\end{aligned}$$

where $\mathcal{F}_x[f](t)$ is the Fourier transform with parameters $a = b = 1$, and the moment-generating function is:

$$\begin{aligned}M(t) &= \int_0^{\infty} \frac{e^{tx} x^{\alpha-1} e^{-x/\theta} dx}{\Gamma(\alpha)\theta^{\alpha}}\\ &= \int_0^{\infty} \frac{x^{\alpha-1}\ e^{-(1-\theta t)x/\theta} dx}{\Gamma(\alpha)\theta^{\alpha}}.\end{aligned}$$

giving moments about 0 of,

$$\mu'_r = \frac{\theta' \Gamma(\alpha+r)}{\Gamma(\alpha)}$$

In order to explicitly find the moments of the distribution using the moment-generating function, let,

$$y \equiv \frac{(1-\theta t)x}{\theta}$$

$$dy = \frac{1-\theta t}{\theta} dx,$$

so,

$$M(t)=\int_0^{\infty}\left(\frac{\theta y}{1-\theta t}\right)^{\alpha-1}\frac{e^{-y}}{\Gamma(\alpha)\theta^{\alpha}}\frac{\theta dy}{1-\theta t}$$

$$=\frac{1}{(1-\theta t)^{\alpha}\Gamma(\alpha)}\int_0^{\infty}y^{\alpha-1}e^{-y}\,dy=\frac{1}{(1-\theta\,t)^{\alpha}},$$

giving the logarithmic moment-generating function as:

$$R(t)=-\alpha\,\mathrm{In}(1-\theta t)$$

$$R'(t)=\frac{\alpha\theta}{1-\theta t}$$

$$R''(t)=\frac{\alpha\theta^2}{(1-\theta t)^2}.$$

The mean, variance, skewness, and kurtosis excess are then,

$$\mu=\alpha\theta$$

$$\sigma^2=\alpha\theta^2$$

$$\gamma_1=\frac{2}{\sqrt{\alpha}}$$

$$\gamma_2=\frac{6}{\alpha}.$$

The gamma distribution is closely related to other statistical distributions. If $X_1, X_2, ..., X_n$ are independent random variates with a gamma distribution having parameters $(\alpha_1,\theta),(\alpha_2,\theta),...,(\alpha_n,\theta)$, then $\sum_{i=1}^{n}X_i$ is distributed as gamma with parameters:

$$\alpha=\sum_{i=1}^{n}\alpha_i$$

$$\theta=\theta.$$

Also, if X_1 and X_2 are independent random variates with a gamma distribution having parameters (α_1,θ) and (α_2,θ), then $X_1/(X_1+X_2)$ is a beta distribution variate with parameters (α_1,α_2). Both can be derived as follows:

$$P(x_1,x_2)=\frac{1}{\Gamma(\alpha_1)\Gamma(\alpha_2)}e^{x_1+x_2}x_1^{\alpha_1-1}x_2^{\alpha_2-1}$$

Let,

$$u=x_1+x_2\quad x_1=u\,v$$

$$v=\frac{x_1}{x_1+x_2}\quad x_2=u(1-v),$$

Then the Jacobian is:

$$J\left(\frac{x_1, x_2}{u, v}\right) = \begin{vmatrix} v & u \\ 1-v & -u \end{vmatrix} = -u,$$

so,

$$g(u,v)\,d v\,d v = f(x,y)\,d x\,d y = f(x,y)\,u\,d u\,d v.$$

$$g(u,v) = \frac{u}{\Gamma(\alpha_1)\Gamma(\alpha_2)} e^{-u} (u v)^{\alpha_1 - 1} u^{\alpha_2 - 1} (1-v)^{\alpha_2 - 1}$$

$$= \frac{1}{\Gamma(\alpha_1)\Gamma(\alpha_2)} e^{-u} u^{\alpha_1 + \alpha_2 - 1} v^{\alpha_1 - 1} (1-v)^{\alpha_2 - 1}.$$

The sum $X_1 + X_2$ therefore has the distribution,

$$f(u) = f(x_1 + x_2) = \int_0^1 g(u,v)\,d v = \frac{e^{-u} u^{\alpha_1 + \alpha_2 - 1}}{\Gamma(\alpha_1 + \alpha_2)},$$

which is a gamma distribution, and the ratio $X_1 / (X_1 + X_2)$ has the distribution,

$$h(v) = h\left(\frac{x_1}{x_1 + x_2}\right)$$

$$= \int_0^\infty g(u,v)\,d u$$

$$= \frac{v^{\alpha_1 - 1}(1-v)^{\alpha_2 - 1}}{B(\alpha_1, \alpha_2)},$$

Where B is the beta function, which is a beta distribution.

If X and Y are gamma variates with parameters α_1 and α_2, the X/Y is a variate with a beta prime distribution with parameters α_1 and α_2. Let,

$$u = x + y \quad v = \frac{x}{y},$$

then the Jacobian is,

$$J\left(\frac{u, v}{x, y}\right) = \begin{vmatrix} 1 & 1 \\ \frac{1}{y} & -\frac{x}{y^2} \end{vmatrix} = -\frac{x+y}{y^2} = -\frac{(1+v)^2}{u},$$

so,

$$d x\,d y = \frac{u}{(1-v)^2} d u\,d v$$

$$g(u,v)=\frac{1}{\Gamma(\alpha_1)\Gamma(\alpha_2)}e^{-u}\left(\frac{u\,v}{1+v}\right)^{\alpha_1-1}\left(\frac{u}{1+v}\right)^{\alpha_2-1}\frac{u}{(1-v)^2}$$

$$=\frac{1}{\Gamma(\alpha_1)\Gamma(\alpha_2)}e^{-u}u^{\alpha_1+\alpha_2-1}\,v^{\alpha_1-1}(1+v)^{-\alpha_1-\alpha_2}.$$

The ratio X/Y therefore has the distribution,

$$h(v)=\int_0^\infty g(u,v)du=\frac{v^{\alpha_1-1}(1-v)^{-\alpha_1-\alpha_2}}{B(\alpha_1,\alpha_2)},$$

which is a beta prime distribution with parameters (α_1, α_2).

The "standard form" of the gamma distribution is given by letting $y \equiv x/\theta$, so $dy = dx/\theta$ and,

$$P(y)dy=\frac{x^{\alpha-1}e^{-x/\theta}}{\Gamma(\alpha)\theta^\alpha}dx$$

$$=\frac{(\theta y)^{\alpha-1}e^{-y}}{\Gamma(\alpha)\theta^\alpha}(\theta dy)$$

$$=\frac{y^{\alpha-1}e^{-y}}{\Gamma(\alpha)}dy,$$

so the moments about o are,

$$v_r=\frac{1}{\Gamma(\alpha)}\int_0^\infty e^{-x}x^{\alpha-1+r}dx$$

$$=\frac{\Gamma(\alpha+r)}{\Gamma(\alpha)}$$

$$=(\alpha)_r,$$

where $(\alpha)_r$ is the Pochhammer symbol. The moments about $\mu=\mu_1$ are then,

$$\mu_1=\alpha$$

$$\mu_2=\alpha$$

$$\mu_3=2\alpha$$

$$\mu_4=3\alpha^2+6\alpha.$$

The moment-generating function is:

$$M(t)=\frac{1}{(1-t)^\alpha},$$

and the cumulant-generating function is:

$$K(t)=\alpha\ \text{In}\ (1-t)=\alpha\left(t+\frac{1}{2}t^2+\frac{1}{3}t^3+\ldots\right),$$

so the cumulants are,

$$k_r = \alpha \Gamma(r).$$

If X is a normal variate with mean μ and standard deviation σ, then,

$$Y \equiv \frac{(X-\mu)^2}{2\sigma^2}$$

is a standard gamma variate with parameter $\alpha = 1/2$.

Cauchy Distribution

The Cauchy distribution, also called the Lorentzian distribution or Lorentz distribution, is a continuous distribution describing resonance behavior. It also describes the distribution of horizontal distances at which a line segment tilted at a random angle cuts the x-axis.

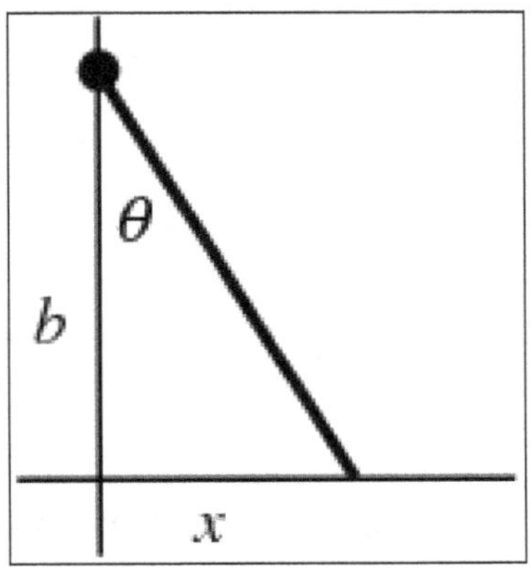

Let θ represent the angle that a line, with fixed point of rotation, makes with the vertical axis,

Then,

$$\tan\theta = \frac{x}{b}$$

$$\theta = \tan^{-1}\left(\frac{x}{b}\right)$$

$$d\theta = \frac{1}{1+\frac{x^2}{b^2}}\frac{dx}{b}$$

$$= \frac{b\,dx}{b^2 + x^2},$$

so the distribution of angle θ is given by,

$$\frac{d\theta}{\pi} = \frac{1}{\pi}\frac{b\,dx}{b^2 + x^2}.$$

This is normalized over all angles, since, $\int_{-\pi/2}^{\pi/2} \frac{d\theta}{\pi} = 1$ and,

$$\int_{-\infty}^{\infty} \frac{1}{\pi} \frac{b\,dx}{b^2 + x^2} = \frac{1}{\pi}\left[\tan^{-1}\left(\frac{x}{b}\right)\right]_{-\infty}^{\infty}$$
$$= \frac{1}{\pi}\left[\frac{1}{2}\pi - \left(-\frac{1}{2}\pi\right)\right]$$
$$= 1.$$

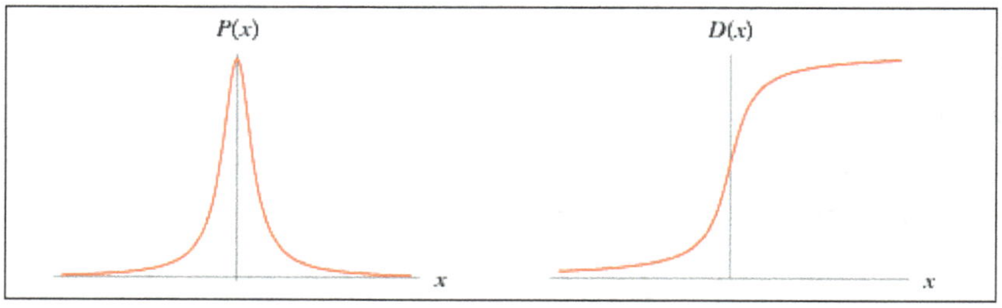

The general Cauchy distribution and its cumulative distribution can be written as:

$$P(x) = \frac{1}{\pi} \frac{b}{(x-m)^2 + b^2}$$
$$D(x) = \frac{1}{2} + \frac{1}{\pi} \tan^{-1}\left(\frac{x-m}{b}\right),$$

Where, b is the half width at half maximum and m is the statistical median. In the illustration about, m = 0.

The Cauchy distribution is implemented in the Wolfram Language as CauchyDistribution[m, Gamma/2].

The characteristic function is:

$$\phi(t) = \frac{1}{\pi} \int_{-\infty}^{\infty} e^{itx} \frac{\frac{1}{2}\Gamma}{\left(\frac{1}{2}\Gamma\right)^2 + (x-m)^2} dx$$
$$= e^{imt - \Gamma|r|/2}.$$

The moments μ_n of the distribution are undefined since the integrals,

$$\mu_n = \int_{-\infty}^{\infty} \frac{\Gamma}{2\pi} \frac{x^n}{(x-m)^2 + \left(\frac{1}{2}\Gamma\right)^2} dx$$

diverge for $n \geq 1$.

If X and Y are variates with a normal distribution, then $Z \equiv X/Y$ has a Cauchy distribution with statistical median $m = 0$ and full width,

$$\Gamma = \frac{2\sigma_x}{\sigma_y}.$$

The sum of n variates each from a Cauchy distribution has itself a Cauchy distribution, as can be seen from,

$$P_n(x) = \mathcal{F}_t^{-1}\left\{\left[\phi(t)\right]^n\right\}(x)$$

$$= \frac{\left(\frac{1}{2}n\Gamma\right)}{\pi\left[\left(\frac{1}{2}nT\right)^2 + (x - nm)^2\right]}.$$

Sampling and Confidence Levels

Sampling Level

The term population is used to refer different meanings in different contexts. For example, consider birthweights of newborns in a particular city over an year period. There were 10,000 childbirths in that particular city. Here the population consists of all those 10,000 newborns. Instead of measuring the weights of all those newborns (which is not challenging in this case), let's decide to measure birthweights of only 50 random newborns. This subset is what is known in statistics as sample. We can use sample as a proxy for entire population and extrapolate measurements done in sample to make conclusions about the populations. In biomedical research population is not only a much larger dataset consisting of target people but also the future generations, therefore we assume that it is infinite. In experimental research, population means the ideal situation or underlying mechanism. For example, Gregor Mendel studied 929 pea plants in F2 generation (his sample) to arrive in generalized 3:1 phenotypic ratio (dominant: recessive) in ideal population. His phenotypic ratio of 3:1 is a model- a mathematical description of a simplified view of nature. He used his samples to extrapolate an underlying phenomenon (his model) that is applicable to diploid organisms in general. Statistics goes from sample (specific things) to population (general conclusion and model) and therefore it is considered as an example of logical induction. On the other hand, probability is an example of logical deduction as it goes other way around, from general to specific.

There are several reasons in statistics why we sample instead of analyzing the whole of a population. Studying a whole population is often cumbersome and resource extensive (expensive!). It may not be possible to identify and measure each individual in the population, or it may be destructive to do so. Luckily, the mathematics of statistics (probability!) allows us to take a sample from a population and make inferences to a population. Simple random sampling is one of the core concepts to much of data collection and analysis. In simple random sampling, each individual or object in a population has an equal probability of being selected into the sample. Often times the first step in simple random sampling is to define the population of concern (often called a sampling frame). Individuals or objects are then selected through a random process from this population (often using a random number generator). Other methods of sampling include systematic random sampling, cluster sampling and stratified random sampling.

Sampling Error and the Sampling Distribution of the Sample Mean

Because we sample (at random), the distribution of each sample we collect will differ from the population (and each other) due to the random processes of sampling. Our samples will not exactly mimic our population of concern. The difference between a sample statistic (such as a mean, xbar) and the true population parameter (such as mu), is called the SAMPLING ERROR. We can develop a sampling distribution of the sample means to see the distribution of means of multiple samples. For example, from a population of incoming MEM students, we might sample 50 students and ask them their age. Our first sample gives us a mean of 25.1. We then repeat this 99 more times (100 different samples, each with a sample size of 50).

We can then develop a histogram of the sample means! This is what is called the 'sampling distribution of the sample mean'. We would expect the distribution of sample means to be less dispersed than the distribution of the ages of ALL incoming MEM students. If the sample size is large enough, we would expect than the mean of the sample means would approach the true population mean. In addition, as the sample size increases, the distribution of the means will approach the normal distribution. With a large sample size, the sample means are normally distributed with a mean of μ (mu) and a standard deviation of σ/sqrt(n). The standard deviation of the sample means is called the standard error of the mean (σ/sqrt(n)). In statistical notation:

$$\bar{X} \sim N\left(\mu, \frac{\sigma}{\sqrt{n}}\right)$$

SE vs. SD

What is the difference between standard deviation and standard error?

- SD is the typical deviation from the average; it doesn't depend on random sampling.
- SE is the typical deviation from the expected value in a random sample. SE results from random sampling.

The profundity of the Central Limit Theorem: As sample size gets larger, even if you start with a non-normal distribution, the sampling distribution approaches a normal distribution.

The Central Limit Theorem

The essence of the Central Limit Theorem: As the sample size increases, the sampling distribution of the sample mean (xbar) concentrates more and more around μ (the population mean). The shape of the distribution also gets closer and closer to the normal distribution as sample size n increases. An example follows. The figure below shows a histogram of our (make believe) population.

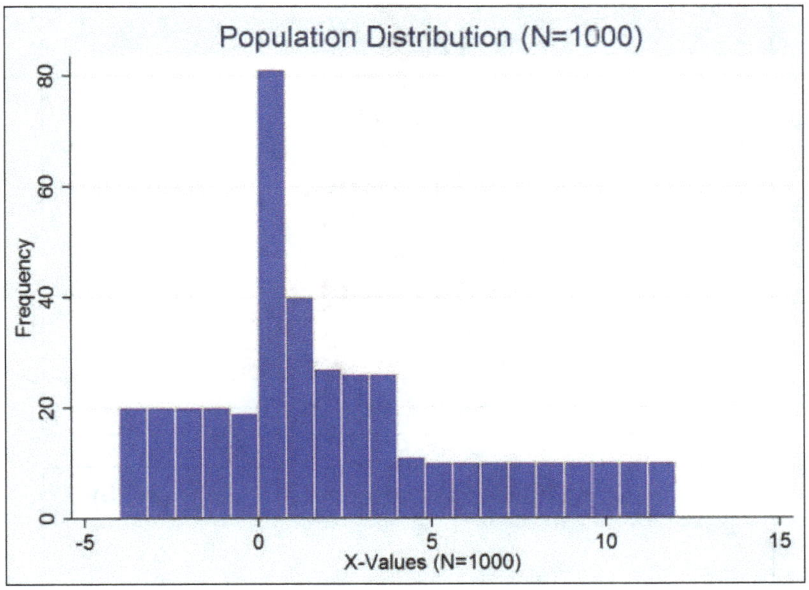

Typically the exact distribution of a population is UNKNOWN, but to demonstrate the Central Limit Theorem, we will start with this known distribution (in blue). This population has a mean (mu) of 2.25 and a standard deviation of 3.93. From this population distribution, I randomly selected a sample of 2 (n=2) and calculated an average (xbar). I then repeated this for a total of 1,000 times and made a histogram of the 1,000 sample means.

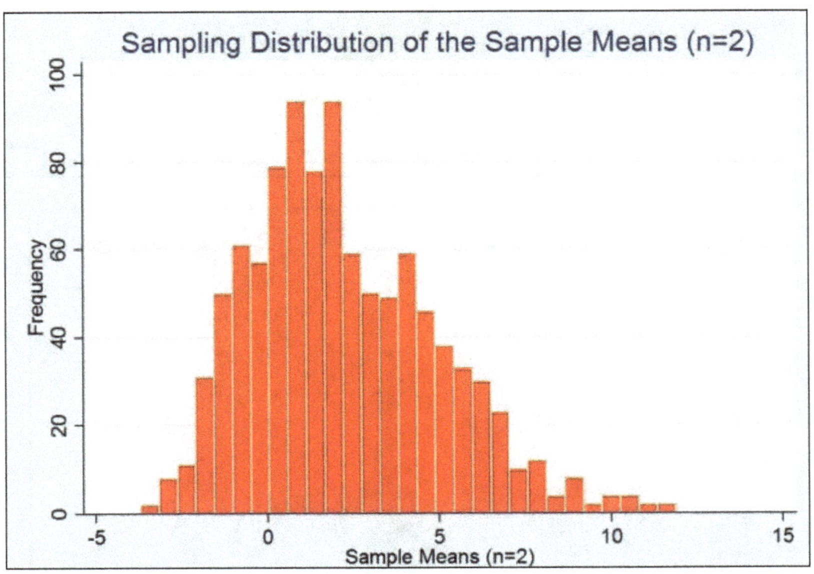

As you can see in the red histogram (sample size n=2), the dispersion of the distribution of sample means is less than the parent population (a greater concentration of values around the mean). The empirical mean of this distribution is 2.31 with a standard deviation of 2.79. As you can see, as sample size increases, the distribution gets increasingly narrow and increasingly approaches a normal distribution. This is the essence of the Central Limit Theorem.

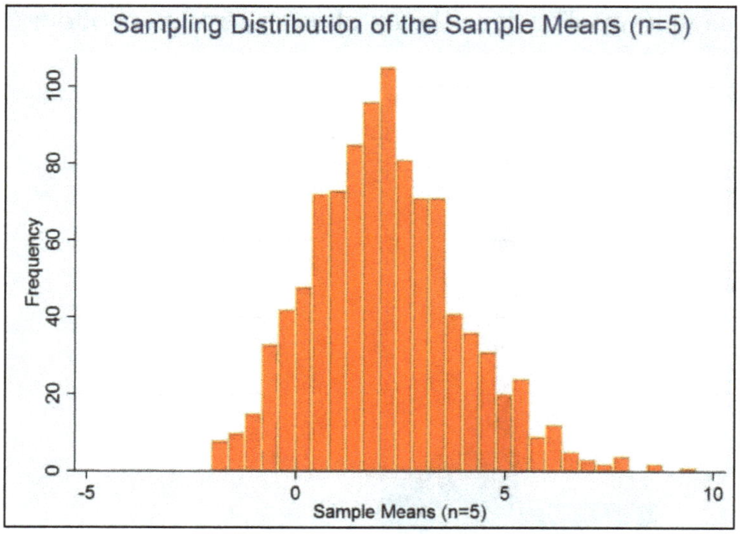
Sampling Distribution of the Sample Means (n=5)
Frequency
0
20
40
60
80
100
-5
0
5
10
Sample Means (n=5)

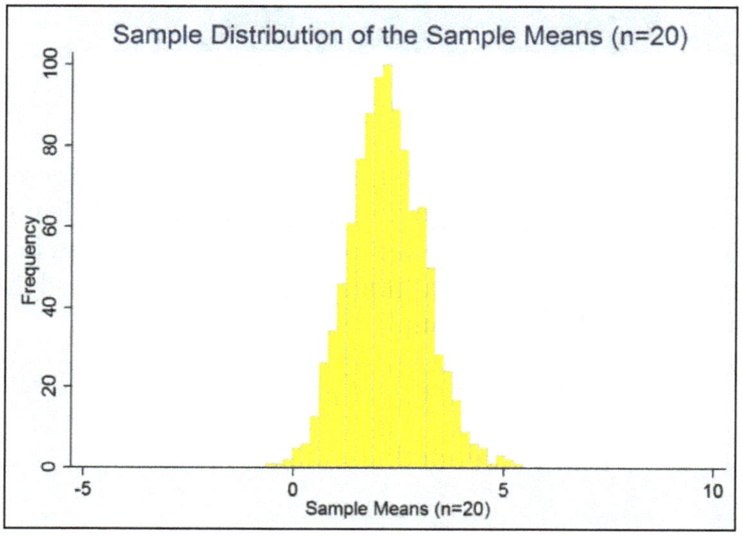
Sample Distribution of the Sample Means (n=20)
Frequency
0
20
40
60
80
100
-5
0
5
10
Sample Means (n=20)

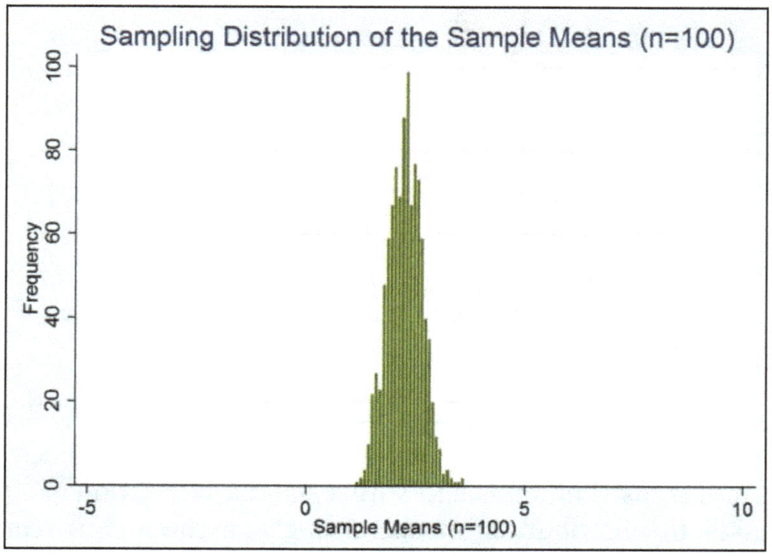
Sampling Distribution of the Sample Means (n=100)
Frequency
0
20
40
60
80
100
-5
0
5
10
Sample Means (n=100)

Calculating Z-Scores with the Sampling Distribution of the Sample Means

$$Z = \frac{\bar{X} - \mu}{\sigma / \sqrt{n}}$$

The formula to calculate a z-score based on a sample mean is listed above. The z-score of a sample mean equals the sample mean minus the population mean, all divided by the standard error of the sample mean. Remember, when dealing with sampling, and therefore sampling error, you need to use the standard error in you z-calculations.

Example: Let's say we have a production process that produces long-lasting light bulbs. The average lifespan of the bulbs produced is 1,500 hours with a population standard deviation of 300 hours.

- We select 100 light bulbs at random. What is the standard deviation of the sample means?
- What is the probability that one bulb, selected at random, will last longer than 1,800 hours?
- What is the probability that the average of 100 randomly selected bulbs is greater than 1,800 hours?

Solution:

- The standard deviation of the sample means equals the known population standard deviation divided by the square root of the sample size (n). Therefore the SD(xbar)= 300/sqrt(100) = 300/10= 30 hours.
- Because we are interested in just one bulb (not an average), we use this z-score formula:

$$Z = \frac{X - \mu}{\sigma}$$

Therefore, Z=(1,800 – 1,500)/300 = 300/300 = 1. And we want to know the probability that X>1,800 which translates into Z>1. We look this up in our z table and find that p(Z>1) = 1- 0.8413=0.1587. There is approximately 15.9% chance that one light bulb chosen at random will last longer than 1,800 hours.

- Because we are interested in the average, we use the following z-score formula:

$$Z = \frac{\bar{X} - \mu}{\sigma / \sqrt{n}}$$

Therefore, z=(1,800 – 1,500)/(300/sqrt(100))=300/30 = 10. And the probability that z>10 is practically zero.

Confidence Level

Let's consider our former example. We precisely know the birth weights of 10,000 newborns in a particular city over a year and let's consider it as our population. We have also calculated mean birth weight, 2.7kg (population mean). Out of 10,000 newborns, we have randomly selected 50 newborns (our sample) and calculated the sample mean. The value of sample mean could be more or less than the actual population mean. Let sample mean be 2.6 kg with 0.1 kg as standard deviation. We will first calculate a statistic called t-ratio from this:

$$t = (m - \mu)/(s/\sqrt{n})$$

Where, m is sample mean, μ is population mean and denominator (s/√n) is sample SEM.

t-ratio can be defined as difference between sample mean and population mean upon sample standard error or the mean.

$$t = (2.6 - 2.7)/(0.1/\sqrt{50})$$
$$= -0.1/0.141$$
$$= -0.70922$$

The value of t ratio is a bit less than zero. This is expected, as the sample mean will be more or less close to population mean, numerator of t ratio will be close to zero, so the ratio will hover around zero. Now we take yet another 50 random samples from the same population and calculate t-ratio. We repeat this 100 times and plot the distribution of t-ratio, like how we plot a histogram. The distribution would look like image below:

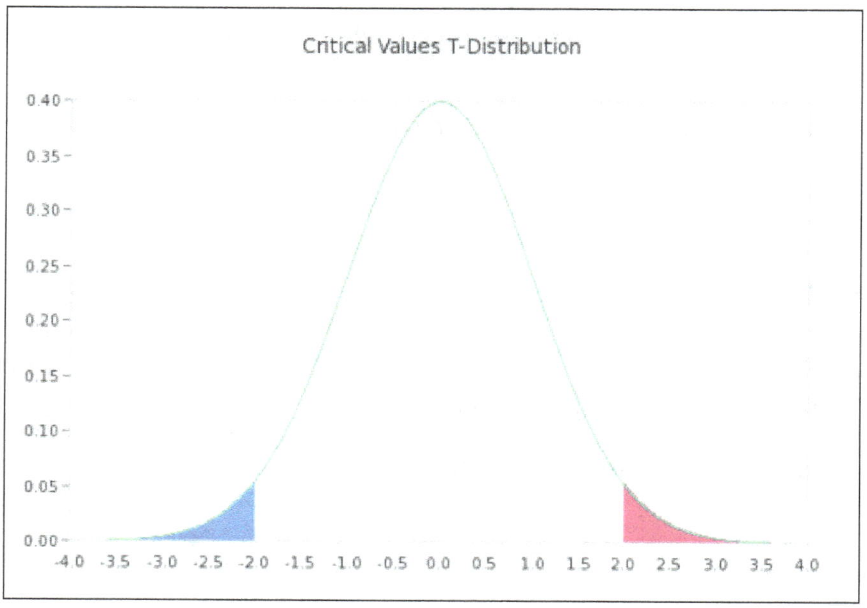

This is one way to convert approximate Gaussian distribution to a symmetric distribution, but for this method, you should know the true population mean (true population mean remains unknown mostly, except in this kind of simulation studies). As explained, t-ratio will be centered around zero (peak at zero), because most of the sample means would be close to population mean. The shape of t-distribution depends upon degree of freedom (df, which is equal to n-1). If the area under 2.5% of total area at both the tails (i.e., most unusual t-ratios whether it is too low or too high) is chopped off, the resulting area would include range of t ratios that include 95% of samples. To get

95% area, we have to chop the tail precisely at a t score at both the directions; this t score is called t* or t critical. In this case, with degree of freedom 49 and significance level 0.05 (because we have to chop of 5% of most unusual values), t* is 2.01 (calculated using online calculator as explained below). That would mean, if we cut the graph at -2.01 and +2.01 (shaded area in figure) we will get range of t-ratios that include 95% of samples. The shape of t distribution (so as t*) depends upon degree of freedom. For a particular significance level, we can calculate t* from t-distribution table.

$$t \text{ ratio} = (m-\mu)/(s/\sqrt{n})$$

We can rearrange this equation to solve for μ, the true population mean. For the sake of brevity, derivation of following formula is omitted:

$$\mu = m \pm t^*\left(s/\sqrt{n}\right)$$

where m is sample mean, t* is constant from t-distribution and $s/\sqrt{n}$ is sample SEM.

Given the sample mean m, sample standard deviation s and sample size n, it is possible for us to define ranges that include the real population mean with confidence. As already explained, t* depends on the desired confidence, which is arbitrarily 95%. If t* for 95% confidence is used, the resulting ranges of sample means would include the true population mean 95% of times. This range is called 95% Confidence Interval.

Confidence Limits, Levels and Intervals

The desired amount of statistical confidence is called Confidence Level. For eg., if you want a very high confidence on your results, you should choose 99% Confidence Level as part of your experimental design. Confidence Level is chosen before experiment is conducted and it is not ethical to change the CL after the data is generated. If you had chosen 99% CL, you will generate 99% Confidence Interval after the experiment. Confidence Interval is a range (lower limit to upper limit) that plot the precision of your sample measurement in comparison with the true population value. Two values that limit this range, the lower limit and upper limit, are called Confidence Limits.

Confidence Interval of Mean

Confidence interval of the mean tells you how precisely you have determined the sample mean as an estimate of population mean. On the other hand, precision depends on Confidence Level (CL). CL of 99 is more precise estimate than 95, which is better than 90. As precision increases, wider would be Confidence Interval. For example, 99% CI is a lot wider than 90% CI.

Another statistic Standard Error of the Mean is also similar to CI; SEM also tells you how precisely you have determined sample mean comparing with population mean. However confidence level of SEM is very low; only around 60%, so SEM is quite an inaccurate measure to quantify precision. 95% CI is routinely used across biological and environmental sciences.

In situations where whole population is used to calculate statistic, for example mean of population, Confidence Interval makes no sense. Consider a class with total strength 24 students and mean mark 11.8 out of 25. Here 11.8 is the population mean and we are 100% sure (confident) that the true population mean is this value, no question about it. However, out of 24 students, if I randomly selects 8 students and calculate their mean marks, that mean would be sample mean and CI makes sense in such situations.

95% CI of sample mean can be calculated using two methods. The parametric (distribution-dependent) method assume that our sample is sampled from a roughly Gaussian distribution. This method utilizes a constant from t-distribution (t*) for calculating 95% CI of sample mean. We have already explained how this formula is derived in earlier section. Formula is:

$$m \pm t^{*} \left(s/\sqrt{n}\right)$$

where m is sample mean, t* is a constant from t-distribution, s is sample standard deviation and n is sample size. Also note that $s/\sqrt{n}$ = SEM (Standard Error of the Mean).

We can calculate 95% CI of sample mean given sample mean, sample standard deviation and sample size; raw data is not necessary.

Confidence Interval is a numerical presented as sample mean ± CI, written as (Lower Limit to Upper Limit).

E.g. If 95% CI is 6 and μ=51, CI range= (45 to 57). Notation like 51 ± 6, which is commonly used to describe standard deviation, is not used to describe Confidence Intervals.

Let us consider an example for calculating 95%CI of sample mean. Test marks data with Mean=12.81 S=4.905 n=24. As n=24, df is 23.

Let's first look up t* for 95% Confidence Level with df=23.

Table: Critical values of the distribution for computing Cls.

	DESIRED CONFIDENCE LEVEL			
df	80%	90%	95%	99%
1	3.0777	6.3138	12.7062	63.6567
2	1.8856	2.9200	4.3027	9.9248
3	1.6377	2.3534	3.1824	5.8409
4	1.5332	2.1318	2.7764	4.6041
5	1.4759	2.0150	2.5706	4.0321
6	1.4398	1.9432	2.4469	3.7074
7	1.4149	1.8946	2.3646	3.4995
8	1.3968	1.8595	2.3060	3.3554
9	1.3830	1.8331	2.2622	3.2498
10	1.3722	1.8125	2.2281	3.1693
11	1.3634	1.7959	2.2010	3.1058
12	1.3562	1.7823	2.1788	3.0545
13	1.3502	1.7709	2.1604	3.0123
14	1.3450	1.7613	2.1448	2.9768
15	1.3406	1.7531	2.1314	2.9467
16	1.3368	1.7459	2.1199	2.9208
17	1.3334	1.7396	2.1098	2.8982
18	1.3304	1.7341	2.1009	2.8784
19	1.3277	1.7291	2.0930	2.8609
20	1.3253	1.7247	2.0860	2.8453
21	1.3232	1.7207	2.0796	2.8314
22	1.3212	1.7171	2.0739	2.8188
23	1.3195	1.7139	2.0687	2.8073
24	1.3178	1.7109	2.0639	2.7969
25	1.3163	1.7081	2.0595	2.7874
26	1.3150	1.7056	2.0555	2.7787

	DESIRED CONFIDENCE LEVEL			
df	80%	90%	95%	99%
27	1.3137	1.7033	2.0518	2.7707
28	1.3125	1.7011	2.0484	2.7633
29	1.3114	1.6991	2.0452	2.7564
30	1.3104	1.6973	2.0423	2.7500
35	1.3062	1.6896	2.0301	2.7238
40	1.3031	1.6839	2.0211	2.7045
45	1.3006	1.6794	2.0141	2.6896
50	1.2987	1.6759	2.0086	2.6778
55	1.2971	1.6730	2.0040	2.6682
60	1.2958	1.6706	2.0003	2.6603
65	1.2947	1.6686	1.9971	2.6536
70	1.2938	1.6669	1.9944	2.6479
75	1.2929	1.6654	1.9921	2.6430
80	1.2922	1.6641	1.9901	2.6387
85	1.2916	1.6630	1.9883	2.6349
90	1.2910	1.6620	1.9867	2.6316
95	1.2905	1.6611	1.9853	2.6286
100	1.2901	1.6602	1.9840	2.6259
150	1.2872	1.6551	1.9759	2.6090
200	1.2858	1.6525	1.9719	2.6006
250	1.2849	1.6510	1.9695	2.5956
300	1.2844	1.6499	1.9679	2.5923
350	1.2840	1.6492	1.9668	2.5899
400	1.2837	1.6487	1.9659	2.5882
450	1.2834	1.6482	1.9652	2.5868
500	1.2832	1.6479	1.9647	2.5857

When computing a Cl of a mean, df equals (n – 1). More generally, df equals n minus the number of parameters being estimated.

t^* is 2.0687

Let's now calculate w, the width of 95% CI:

$W = t^* (s/\sqrt{n})$

$= 2.0687 * (4.905 / \sqrt{24})$

$= 2.0687 * (4.905 / 4.90)$

$= 2.0687 * 1$

$= 2.0687$

Finally, 95% CI of mean is,

Mean – 2.0687 to Mean + 2.0687

(10.7413 to 14.8787)

In case no Standard Deviation or Mean are given but presented with raw data, we have to calculate mean and SD to calculate 95% CI.

Microsoft Excel formula for t^* in CI formula is:

=TINV (alpha, n – 1)

Where alpha is level of significance (0.05 to calculate 95% CI).

Hypothesis Testing

On the basis of the central limit theorem, we know that the probability of selecting any other sample mean value from this population is normally distributed.

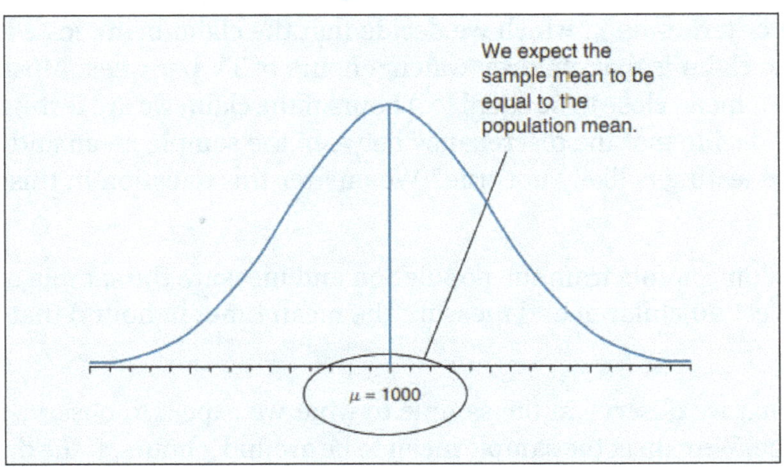

In research, we select samples to learn more about populations of interest to us. In terms of the mean, we measure a sample mean to learn more about the mean in a population. Therefore, we will use the sample mean to describe the population mean. We begin by stating the value of a population mean, and then we select a sample and measure the mean in that sample. On average, the value of the sample mean will equal the population mean. The larger the difference or discrepancy between the sample mean and population mean, the less likely it is that we could have selected that sample mean, if the value of the population mean is correct. This type of experimental situation, using the example of standardized exam scores, is illustrated in figure.

The sampling distribution for a population mean is equal to 1,000. If 1,000 is the correct population mean, then we know that, on average, the sample mean will equal 1,000 (the population mean). Using the empirical rule, we know that about 95% of all samples selected from this population will have a sample mean that falls within two standard deviations (SD) of the mean. It is therefore unlikely (less than a 5% probability) that we will measure a sample mean beyond 2 SD from the population mean, if the population mean is indeed correct.

The method in which we select samples to learn more about characteristics in a given population is called hypothesis testing. Hypothesis testing is really a systematic way to test claims or ideas about a group or population. To illustrate, suppose we read an article stating that children in the United States watch an average of 3 hours of TV per week. To test whether this claim is true, we record the time (in hours) that a group of 20 American children (the sample), among all children in the United States (the population), watch TV. The mean we measure for these 20 children is a sample mean. We can then compare the sample mean we select to the population mean.

Hypothesis testing or significance testing is a method for testing a claim or hypothesis about a parameter in a population, using data measured in a sample. In this method, we test some hypothesis by determining the likelihood that a sample statistic could have been selected, if the hypothesis regarding the population parameter were true.

The method of hypothesis testing can be summarized in four steps:

- To begin, we identify a hypothesis or claim that we feel should be tested. For example, we might want to test the claim that the mean number of hours that children in the United States watch TV is 3 hours.
- We select a criterion upon which we decide that the claim being tested is true or not. For example, the claim is that children watch 3 hours of TV per week. Most samples we select should have a mean close to or equal to 3 hours if the claim we are testing is true. So at what point do we decide that the discrepancy between the sample mean and 3 is so big that the claim we are testing is likely not true? We answer this question in this step of hypothesis testing.
- Select a random sample from the population and measure the sample mean. For example, we could select 20 children and measure the mean time (in hours) that they watch TV per week.
- Compare what we observe in the sample to what we expect to observe if the claim we are testing is true. We expect the sample mean to be around 3 hours. If the discrepancy between

the sample mean and population mean is small, then we will likely decide that the claim we are testing is indeed true. If the discrepancy is too large, then we will likely decide to reject the claim as being not true.

Four Steps to Hypothesis Testing

The goal of hypothesis testing is to determine the likelihood that a population parameter, such as the mean, is likely to be true.

- Step 1: State the hypotheses.
- Step 2: Set the criteria for a decision.
- Step 3: Compute the test statistic.
- Step 4: Make a decision.

Step 1: State the hypotheses. We begin by stating the value of a population mean in a null hypothesis, which we presume is true. For the children watching TV example, we state the null hypothesis that children in the United States watch an average of 3 hours of TV per week. This is a starting point so that we can decide whether this is likely to be true, similar to the presumption of innocence in a courtroom. When a defendant is on trial, the jury starts by assuming that the defendant is innocent. The basis of the decision is to determine whether this assumption is true. Likewise, in hypothesis testing, we start by assuming that the hypothesis or claim we are testing is true. This is stated in the null hypothesis. The basis of the decision is to determine whether this assumption is likely to be true.

The null hypothesis (Ho), stated as the null, is a statement about a population parameter, such as the population mean, that is assumed to be true.

The null hypothesis is a starting point. We will test whether the value stated in the null hypothesis is likely to be true.

Keep in mind that the only reason we are testing the null hypothesis is because we think it is wrong. We state what we think is wrong about the null hypothesis in an alternative hypothesis. For the children watching TV example, we may have reason to believe that children watch more than (>) or less than (<) 3 hours of TV per week. When we are uncertain of the direction, we can state that the value in the null hypothesis is not equal to (≠) 3 hours.

In a courtroom, since the defendant is assumed to be innocent (this is the null hypothesis so to speak), the burden is on a prosecutor to conduct a trial to show evidence that the defendant is not innocent. In a similar way, we assume the null hypothesis is true, placing the burden on the researcher to conduct a study to show evidence that the null hypothesis is unlikely to be true. Regardless, we always make a decision about the null hypothesis (that it is likely or unlikely to be true). The alternative hypothesis is needed for Step 2.

An alternative hypothesis (H_1) is a statement that directly contradicts a null hypothesis by stating that that the actual value of a population parameter is less than, greater than, or not equal to the value stated in the null hypothesis.

The alternative hypothesis states what we think is wrong about the null hypothesis, which is needed for Step 2.

Making Sense: Testing the Null Hypothesis

A decision made in hypothesis testing centers on the null hypothesis. This means two things in terms of making a decision:

- Decisions are made about the null hypothesis: Using the courtroom analogy, a jury decides whether a defendant is guilty or not guilty. The jury does not make a decision of guilty or innocent because the defendant is assumed to be innocent. All evidence presented in a trial is to show that a defendant is guilty. The evidence either shows guilt (decision: guilty) or does not (decision: not guilty). In a similar way, the null hypothesis is assumed to be correct. A researcher conducts a study showing evidence that this assumption is unlikely (we reject the null hypothesis) or fails to do so (we retain the null hypothesis).
- The bias is to do nothing: Using the courtroom analogy, for the same reason the courts would rather let the guilty go free than send the innocent to prison, researchers would rather do nothing (accept previous notions of truth stated by a null hypothesis) than make statements that are not correct. For this reason, we assume the null hypothesis is correct, thereby placing the burden on the researcher to demonstrate that the null hypothesis is not likely to be correct.

Step 2: Set the criteria for a decision. To set the criteria for a decision, we state the level of significance for a test. This is similar to the criterion that jurors use in a criminal trial. Jurors decide whether the evidence presented shows guilt beyond a reasonable doubt (this is the criterion). Likewise, in hypothesis testing, we collect data to show that the null hypothesis is not true, based on the likelihood of selecting a sample mean from a population (the likelihood is the criterion). The likelihood or level of significance is typically set at 5% in research studies. When the probability of obtaining a sample mean is less than 5% if the null hypothesis were true, then we conclude that the sample we selected is too unlikely and so we reject the null hypothesis.

Level of significance, or significance level, refers to a criterion of judgment upon which a decision is made regarding the value stated in a null hypothesis. The criterion is based on the probability of obtaining a statistic measured in a sample if the value stated in the null hypothesis were true.

In behavioral science, the criterion or level of significance is typically set at 5%. When the probability of obtaining a sample mean is less than 5% if the null hypothesis were true, then we reject the value stated in the null hypothesis.

The alternative hypothesis establishes where to place the level of significance. Remember that we know that the sample mean will equal the population mean on average if the null hypothesis is true. All other possible values of the sample mean are normally distributed (central limit theorem). The empirical rule tells us that at least 95% of all sample means fall within about 2 standard deviations (SD) of the population mean, meaning that there is less than a 5% probability of obtaining a sample mean that is beyond 2 SD from the population mean. For the children watching TV example, we can look for the probability of obtaining a sample mean beyond 2 SD in the upper tail (greater than 3), the lower tail (less than 3), or both tails (not equal to 3). Figure shows that the alternative hypothesis is used to determine which tail or tails to place the level of significance for a hypothesis test.

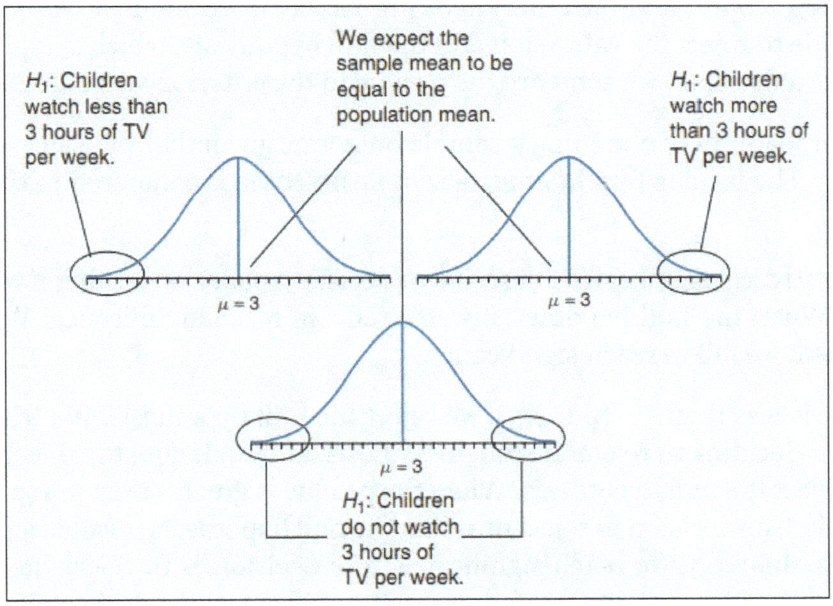

The alternative hypothesis determines whether to place the level of significance in one or both tails of a sampling distribution. Sample means that fall in the tails are unlikely to occur (less than a 5% probability) if the value stated for a population mean in the null hypothesis is true.

Step 3: Compute the test statistic. Suppose we measure a sample mean equal to 4 hours per week that children watch TV. To make a decision, we need to evaluate how likely this sample outcome is, if the population mean stated by the null hypothesis (3 hours per week) is true. We use a test statistic to determine this likelihood. Specifically, a test statistic tells us how far, or how many standard deviations, a sample mean is from the population mean. The larger the value of the test statistic, the further the distance, or number of standard deviations, a sample mean is from the population mean stated in the null hypothesis. The value of the test statistic is used to make a decision in Step 4.

The test statistic is a mathematical formula that allows researchers to determine the likelihood of obtaining sample outcomes if the null hypothesis were true. The value of the test statistic is used to make a decision regarding the null hypothesis.

Step 4: Make a decision. We use the value of the test statistic to make a decision about the null hypothesis. The decision is based on the probability of obtaining a sample mean, given that the value stated in the null hypothesis is true. If the probability of obtaining a sample mean is less than 5% when the null hypothesis is true, then the decision is to reject the null hypothesis. If the probability of obtaining a sample mean is greater than 5% when the null hypothesis is true, then the decision is to retain the null hypothesis. In sum, there are two decisions a researcher can make:

- Reject the null hypothesis: The sample mean is associated with a low probability of occurrence when the null hypothesis is true.
- Retain the null hypothesis: The sample mean is associated with a high probability of occurrence when the null hypothesis is true.

The probability of obtaining a sample mean, given that the value stated in the null hypothesis is true, is stated by the p value. The p value is a probability: It varies between 0 and 1 and can never be negative. In Step 2, we stated the criterion or probability of obtaining a sample mean at which point we will decide to reject the value stated in the null hypothesis, which is typically set at 5% in research. To make a decision, we compare the p value to the criterion we set in Step 2.

A p value is the probability of obtaining a sample outcome, given that the value stated in the null hypothesis is true. The p value for obtaining a sample outcome is compared to the level of significance.

Significance, or statistical significance, describes a decision made concerning a value stated in the null hypothesis. When the null hypothesis is rejected, we reach significance. When the null hypothesis is retained, we fail to reach significance.

When the p value is less than 5% ($p < .05$), we reject the null hypothesis. We will refer to $p < .05$ as the criterion for deciding to reject the null hypothesis, although note that when $p = .05$, the decision is also to reject the null hypothesis. When the p value is greater than 5% ($p > .05$), we retain the null hypothesis. The decision to reject or retain the null hypothesis is called significance. When the p value is less than .05, we reach significance; the decision is to reject the null hypothesis. When the p value is greater than .05, we fail to reach significance; the decision is to retain the null hypothesis. Figure shows the four steps of hypothesis testing.

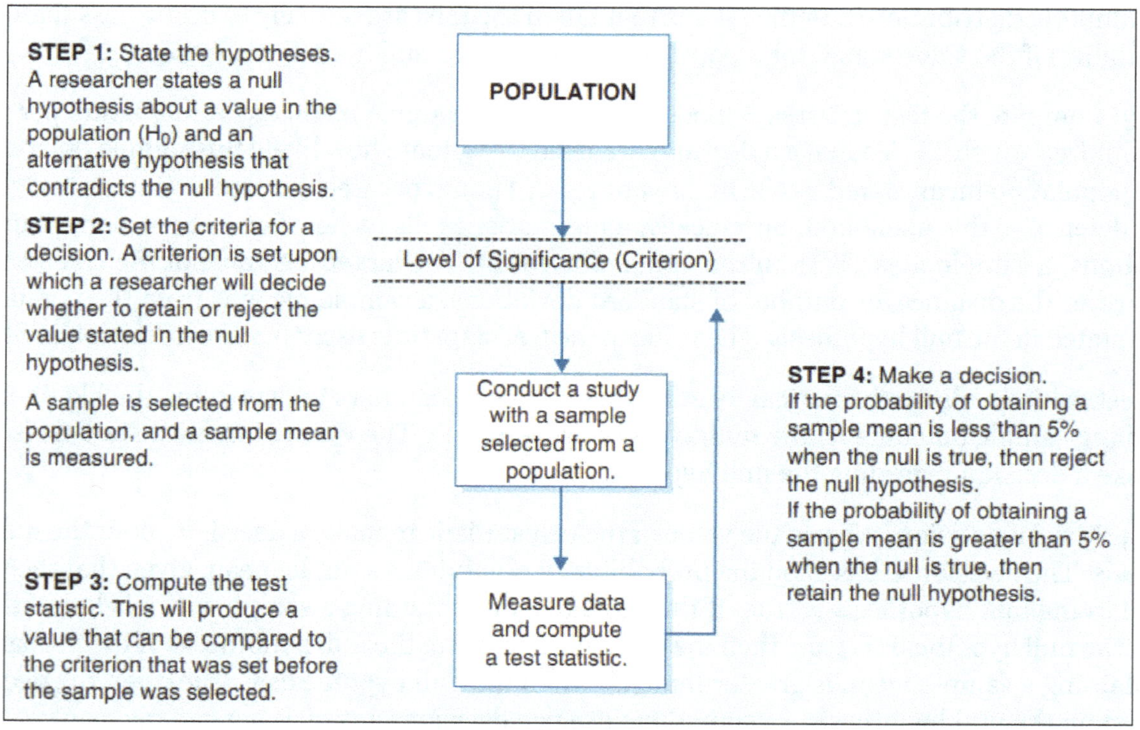

A summary of hypothesis testing.

Decision Making

In Step 4, we decide whether to retain or reject the null hypothesis. Because we are observing a sample and not an entire population, it is possible that a conclusion may be wrong. Table shows that there are four decision alternatives regarding the truth and falsity of the decision we make about a null hypothesis:

- The decision to retain the null hypothesis could be correct.
- The decision to retain the null hypothesis could be incorrect.
- The decision to reject the null hypothesis could be correct.
- The decision to reject the null hypothesis could be incorrect.

Table: Four outcomes for making a decision. The decision can be either correct (correctly reject or retain null) or wrong (incorrectly reject or retain null).

		Decision	
		Retain the null	Reject the null
Truth in the population	True	Correct $1-\alpha$	Type I Error α
	False	Type II Error β	Correct $1-\beta$ Power

We investigate each decision alternative here. Since we will observe a sample, and not a population, it is impossible to know for sure the truth in the population. So for the sake of illustration, we will assume we know this. This assumption is labeled as truth in the population in table.

Decision: Retain the Null Hypothesis

When we decide to retain the null hypothesis, we can be correct or incorrect. The correct decision is to retain a true null hypothesis. This decision is called a null result or null finding. This is usually an uninteresting decision because the decision is to retain what we already assumed: that the value stated in the null hypothesis is correct. For this reason, null results alone are rarely published in behavioral research.

The incorrect decision is to retain a false null hypothesis. This decision is an example of a Type II error, or b error. With each test we make, there is always some probability that the decision could be a Type II error. In this decision, we decide to retain previous notions of truth that are in fact false. While it's an error, we still did nothing; we retained the null hypothesis. We can always go back and conduct more studies.

Type II error, or beta (b) error, is the probability of retaining a null hypothesis that is actually false.

Decision: Reject the Null Hypothesis

When we decide to reject the null hypothesis, we can be correct or incorrect. The incorrect decision is to reject a true null hypothesis. This decision is an example of a Type I error. With each test we make, there is always some probability that our decision is a Type I error. A researcher who makes this error decides to reject previous notions of truth that are in fact true. Making this type of error is analogous to finding an innocent person guilty. To minimize this error, we assume a defendant is innocent when beginning a trial. Similarly, to minimize making a Type I error, we assume the null hypothesis is true when beginning a hypothesis test.

Type I error is the probability of rejecting a null hypothesis that is actually true. Researchers directly control for the probability of committing this type of error.

An alpha (a) level is the level of significance or criterion for a hypothesis test. It is the largest probability of committing a Type I error that we will allow and still decide to reject the null hypothesis.

Since we assume the null hypothesis is true, we control for Type I error by stating a level of significance. The level we set, called the alpha level (symbolized as α), is the largest probability of committing a Type I error that we will allow and still decide to reject the null hypothesis. This criterion is usually set at .05 ($\alpha = .05$), and we compare the alpha level to the p value. When the probability of a Type I error is less than 5% ($p < .05$), we decide to reject the null hypothesis; otherwise, we retain the null hypothesis.

The correct decision is to reject a false null hypothesis. There is always some probability that we decide that the null hypothesis is false when it is indeed false. This decision is called the power of the decision-making process. It is called power because it is the decision we aim for. Remember that we are only testing the null hypothesis because we think it is wrong. Deciding to reject a false null hypothesis, then, is the power, inasmuch as we learn the most about populations when we accurately reject false notions of truth. This decision is the most published result in behavioral research.

The power in hypothesis testing is the probability of rejecting a false null hypothesis. Specifically, it is the probability that a randomly selected sample will show that the null hypothesis is false when the null hypothesis is indeed false.

P-hacking

Remember the coin toss experiment we described in the beginning to test how lucky a person is? Two ways the person can cheat are by changing the threshold value and increasing the no of trials. Similarly, experimental scientists unethically resort to many practices to decrease the P value (a low P value is desirable, as a P value lesser than the alpha ($P<0.05$) would reject the null hypothesis and lead to a conclusion of statistical significance. Some of these tricks include:

- Try different tests, parametric and nonparametric and picking the one with lowest P value The statistical test that will be used (for example, unpaired t-test) should be clearly specified as part of the experimental design. It is not fair to try different tests and go with the one that gives lowest P value.

- Dynamic sample size and stop when P is lower than threshold Sample size should clearly be decided as part of the experimental design. It is not fair to change this till P value is less than the threshold.

- Slice and dice the dataset to get lowest P Obviously an unethical practice. We should use all of the elements of our dataset without slicing and dicing.

- Cherry-picking the dataset to get lowest P Obviously an unethical practice. We should use all of the elements of our dataset without picking certain values that affirms our preconceived conclusions.

- Play with outliers, remove one or more selectively to get desired P Obviously an unethical practice. We should use all of the elements of our dataset without removing any outliers manually. In case outliers need to be removed, a formal statistic test such as Grubb's test can be used for this.

All of the above are cheating, and construes scientific misconduct. As already mentioned, statistical hypothesis testing based on P values is based upon the threshold significance level, which is nothing but an arbitrary value. Statistical significance is usually denoted by the symbol * in graphs. Over emphasizing the statistical significance and * is sometimes referred as 'stargazing'. P values depend largely on sample size. It is easy to get very low P values with huge sample size with tiny effect, or tiny sample size with huge effect. P value answers the question, is there a significant evidence of difference? That question is different from "Is there an evidence of significant difference?". P value does not answer that question. Difference between sample means might be tiny, or scientifically negligible, yet decision based on P value could be 'statistically significant'. Add on to the woes, many studies have revealed that P values are not very well reproducible. Instead of P values, the current consensus among the statisticians is to used 95% Confidence Intervals.

P-values and Confidence Intervals

Confidence Intervals and P values are very well connected. A 95% Confidence Interval is similar to 0.05 significance level as already explained. The question is does the range of values defined by 95% Confidence Interval include our null hypothesis? For example, while comparing the difference in mean between two groups. Null hypothesis is that the group means are same, or the difference between two group means is zero. If 95% Confidence Intervals of difference between sample means does not include zero-our null Hypothesis, then P0.05) and indicate a statistically significant difference. Let us consider another example. A random sample consisting of 10 individuals volunteered to have their body temperatures measured to know whether they have normal body temperature or not. Mean body temperature was found to be 36.8°C. The width of 95% Confidence Interval was found to be 0.3°C and therefore 95% CI about the sample mean ranges from (36.8-0.3) to (36.8+0.3) that is 36.5°C to 37.1°C. Our null hypothesis of having no fever is 37°C. Does our obtained 95% CI include the null hypothesis of 37°C? Yes, it does. So, P>0.05 and we can conclude that there is no statistically significant difference from the null hypothesis.

FDR and Statistical Power

	Decision: Reject Null Hypothesis	Decision: Do Not Reject Null Hypothesis
Reality: Null Hypothesis is True	Type I Error (False Positive) [A]	(no error) [B]
Reality: Null Hypothesis is False	(no error) [C]	Type II Error (False Negative) [D]

We have already defined A/(A+B) as α, level of significance. It is 'Fraction of false positives when null hypothesis is true'. A related term is False Discovery Rate (FDR), which is A/A+C. FDR is defined as 'Fraction of false positives out of statistically significant results'. If the result is statistically significant, what is the probability that null hypothesis is really true? Note that for both level of significance and FDR, numerator remains same,' fraction of false positives', but denominator is different. While significance level is based out of all results where null hypothesis is true (i.e., all results that in reality have no effect/difference etc.), FDR is based out of results with a 'statistically significant' conclusion. FDR depends upon 'prior probability'-the context of experiment akin to Bayesian Statistics. FDR and Prior Probability will be elaborated in the module of Bayesian Statistics. Another related term is 'statistical power' which is C/C+D. It is 'fraction of statistically significant results out of all results where null hypothesis is false'. The statistical power is the probability to obtain a statistically significant result assuming a certain effect size in population. Power depends on sample size, variability, choice of α and hypothetical effect size. A high power is obtained when: 1) Large sample size is used, 2) Looking for a large effect (or large difference or a strong correlation etc), and 3) Data with little scatter (small variance and SD).

References

- Discrete-Probability: amsi.org.au, Retrieved 06, April 2020
- Continuous-probability-distributions, stats-review: nicholas.duke.edu, Retrieved 25, June 2020
- Gamma-Distribution: wolfram.com, Retrieved 03, August 2020
- Cauchy-Distribution: wolfram.com, Retrieved 16, May 2020
- Stats-review, sampling: nicholas.duke.edu, Retrieved 20, February 2020

4

Determinants and Matrices

Determinants and matrices are used in linear algebra to non-homogenous equations in linear form. Determinants can only be used for square matrices. This chapter has been carefully written to provide an easy understanding of the varied facets of determinants and matrices such as special matrices, adjoint and inverse of a matrix, and rank of a matrix.

Introduction to Matrices

In many economic analysis, variables are assumed to be related by sets of linear equations. Matrix algebra provides a clear and concise notation for the formulation and solution of such problems, many of which would be complicated in conventional algebraic notation. The concept of determinant and is based on that of matrix. Hence we shall first explain a matrix.

Matrix

A set of mn numbers (real or complex), arranged in a rectangular formation (array or table) having m rows and n columns and enclosed by a square bracket [ ] is called $m \times n$ matrix (read "m by n matrix").

An $m \times n$ matrix is expressed as:

$$A = \begin{bmatrix} a_{11} & a_{12} & ----a_{1n} \\ a_{21} & a_{22} & ----a_{2n} \\ --- & ---- & ------ \\ --- & ---- & ------ \\ a_{m1} & a_{m2} & ----a_{mn} \end{bmatrix}$$

The letters a_{ij} stand for real numbers. Note that a_{ij} is the element in the ith row and jth column of the matrix .Thus the matrix A is sometimes denoted by simplified form as (a_{ij}) or by $\{a_{ij}\}$ i.e., $A = (a_{ij})$.

Matrices are usually denoted by capital letters A, B, C etc and its elements by small letters a, b, c etc.

Order of a Matrix

The order or dimension of a matrix is the ordered pair having as first component the number of rows and as second component the number of columns in the matrix. If there are 3 rows and 2 columns in a matrix, then its order is written as (3, 2) or (3 x 2) read as three by two. In general if m are rows and n are columns of a matrix, then its order is (m x n).

Examples:

$$\begin{bmatrix} 1 & 2 & 3 \\ 4 & 5 & 6 \end{bmatrix}, \begin{bmatrix} 1 \\ 2 \\ 3 \end{bmatrix} \text{ and } \begin{bmatrix} a_1 & a_2 & a_3 & a_4 \\ b_1 & b_2 & b_3 & b_4 \\ c_1 & c_2 & c_3 & c_4 \\ d_1 & d_2 & d_3 & d_4 \end{bmatrix}$$

are matrices of orders (2 x 3), (3 x 1) and (4 x 4) respectively.

Some types of Matrices

Row Matrix and Column Matrix

A matrix consisting of a single row is called a row matrix or a row vector, whereas a matrix having single column is called a column matrix or a column vector.

Null or Zero Matrix

A matrix in which each element is '0' is called a Null or Zero matrix. Zero matrices are generally denoted by the symbol O. This distinguishes zero matrix from the real number 0.

For example $0 = \begin{bmatrix} 0 & 0 & 0 & 0 \\ 0 & 0 & 0 & 0 \end{bmatrix}$ is a zero matrix of order 2 x 4.

The matrix O_{mxn} has the property that for every matrix A_{mxn},

$$A + O = O + A = A$$

Square Matrix

A matrix A having same numbers of rows and columns is called a square matrix. A matrix A of order m x n can be written as A_{mxn}. If m = n, then the matrix is said to be a square matrix. A square matrix of order n x n, is simply written as A_n.

Thus $\begin{bmatrix} 2 & 5 \\ 1 & 3 \end{bmatrix}$ and $\begin{bmatrix} a & b & c \\ d & e & f \\ g & h & i \end{bmatrix}$ are square matrix of order 2 and 3.

Main or Principal (Leading) Diagonal

The principal diagonal of a square matrix is the ordered set of elements a_{ij}, where i = j, extending from the upper left-hand corner to the lower right-hand corner of the matrix. Thus, the principal diagonal contains elements a_{11}, a_{22}, a_{33} etc.

For example, the principal diagonal of consists of elements 1, 2 and 0, in that order.

$$\begin{bmatrix} 1 & 3 & -1 \\ 5 & 2 & 3 \\ 6 & 4 & 0 \end{bmatrix}$$

Particular Cases of a Square Matrix

Diagonal Matrix

A square matrix in which all elements are zero except those in the main or principal diagonal is called a diagonal matrix. Some elements of the principal diagonal may be zero but not all.

For example $\begin{bmatrix} 4 & 0 \\ 0 & 2 \end{bmatrix}$ and $\begin{bmatrix} 1 & 0 & 0 \\ 0 & 1 & 0 \\ 0 & 0 & 0 \end{bmatrix}$ are diagonal matrices.

$$\text{In general } A = \begin{bmatrix} a_{11} & a_{12} & --- & a_{1n} \\ a_{21} & a_{22} & --- & a_{2n} \\ --- & --- & --- & --- \\ --- & --- & --- & --- \\ a_{n1} & a_{n2} & --- & a_{nn} \end{bmatrix} = (a_{ij})_{nxn}$$

is a diagonal matrix if and only if:

$$a_{ij} = 0 \qquad \text{for } i \neq j$$

$$a_{ij} \neq 0 \quad \text{for at least one } i = j$$

Scalar Matrix

A diagonal matrix in which all the diagonal elements are same, is called a scalar matrix i.e.

Thus $\begin{bmatrix} 3 & 0 \\ 0 & 3 \end{bmatrix}$ and $\begin{bmatrix} k & 0 & 0 \\ 0 & k & 0 \\ 0 & 0 & k \end{bmatrix}$ are scalar matrices.

Identity Matrix or Unit Matrix

A scalar matrix in which each diagonal element is 1(unity) is called a unit matrix. An identity matrix of order n is denoted by I_n.

Thus $I_2 \begin{bmatrix} 1 & 0 \\ 0 & 1 \end{bmatrix}$ and $I_3 = \begin{bmatrix} 1 & 0 & 0 \\ 0 & 1 & 0 \\ 0 & 0 & 1 \end{bmatrix}$ are the identity matrices of order 2 and 3.

In general, $A = \begin{bmatrix} a_{11} & a_{12} & ----a_{1n} \\ a_{21} & a_{22} & ----a_{2n} \\ --- & ---- & ------ \\ --- & ---- & ------ \\ a_{m1} & a_{m2} & ----a_{mn} \end{bmatrix} = \left[a_{ij}\right]_{mxn}$ is an identity matrix if and only if $a_{ij} = 0$ for $i \neq j$ and $a_{ij} = 1$ for $i = j$.

If a matrix A and identity matrix I are comformable for multiplication, then I has the property that AI = IA = A i.e., I is the identity matrix for multiplication.

Equal Matrices

Two matrices A and B are said to be equal if and only if they have the same order and each element of matrix A is equal to the corresponding element of matrix B i.e for each i, j, $a_{ij} = b_{ij}$:

$$\text{Thus } A = \begin{bmatrix} 2 & 1 \\ 3 & 0 \end{bmatrix} \text{ and } B = \begin{bmatrix} \frac{4}{2} & 2-1 \\ \frac{2}{\sqrt{9}} & 0 \end{bmatrix}$$

Then, A = B because the order of matrices A and B is same and $a_{ij} = b_{ij}$ for every i, j.

Example: Find the values of x, y, z and a which satisfy the matrix equation.

Solution: By the definition of equality of matrices, we have:

$$x+3=0 \quad \ldots\ldots\ldots\ldots(1)$$
$$2y+x=-7 \quad \ldots\ldots\ldots\ldots(2)$$
$$z-1=3 \quad \ldots\ldots\ldots\ldots(3)$$
$$4a-6=2a \quad \ldots\ldots\ldots\ldots(4)$$

From (1) $x=-3$

Put the value of x in (2), we get y = − 2

From (3) z = 4

From (4) a = 3

The Negative of a Matrix

The negative of the matrix A_{mxn}, denoted by $-A_{mxn}$, is the matrix formed by replacing each element in the matrix A_{mxn} with its additive inverse. For example,

$$\text{If } A_{3x2} = \begin{bmatrix} 3 & -1 \\ 2 & -2 \\ -4 & 5 \end{bmatrix}$$

$$\text{Then } -A_{3x2} = \begin{bmatrix} -3 & 1 \\ -2 & 2 \\ 4 & -5 \end{bmatrix}$$

for every matrix A_{mxn}, the matrix $-A_{mxn}$ has the property that,

$$A + (-A) = (-A) + A = 0$$

i.e., (−A) is the additive inverse of A.

The sum $B_{m-n} + (-A_{mxn})$ is called the difference of B_{mxn} and A_{mxn} and is denoted by $B_{mxn} - A_{mxn}$.

Operations on Matrices

Multiplication of a Matrix by a Scalar

If A is a matrix and k is a scalar (constant), then kA is a matrix whose elements are the elements of A, each multiplied by k,

For example, if $A = \begin{bmatrix} 4 & -3 \\ 8 & -2 \\ -1 & 0 \end{bmatrix}$ then for a scalar k,

$$kA = \begin{bmatrix} 4k & -3k \\ 8k & -2k \\ -k & 0 \end{bmatrix}$$

Also, $3\begin{bmatrix} 5 & -8 & 4 \\ 0 & 3 & -5 \\ 3 & -1 & 4 \end{bmatrix} = \begin{bmatrix} 15 & -24 & 12 \\ 0 & 9 & -15 \\ 9 & -3 & 12 \end{bmatrix}$

Addition and Subtraction of Matrices

If A and B are two matrices of same order m×n then their sum A + B is defined as C, m×n matrix such that each element of C is the sum of the corresponding elements of A and B.

For example:

If $A = \begin{bmatrix} 3 & 1 & 2 \\ 2 & 1 & 4 \end{bmatrix}$ and $B = \begin{bmatrix} 1 & 0 & 2 \\ -1 & 3 & 0 \end{bmatrix}$

Then $C = A + B = \begin{bmatrix} 3+1 & 1+0 & 2+2 \\ 2-1 & 1+3 & 4+0 \end{bmatrix} = \begin{bmatrix} 4 & 1 & 4 \\ 1 & 4 & 4 \end{bmatrix}$

Similarly, the difference A − B of the two matrices A and B is a matrix each element of which is obtained by subtracting the elements of B from the corresponding elements of A.

Thus if $A = \begin{bmatrix} 6 & 2 \\ 7 & -5 \end{bmatrix}$, $B = \begin{bmatrix} 8 & 1 \\ 3 & 4 \end{bmatrix}$

Then $A - B = \begin{bmatrix} 6 & 2 \\ 7 & -5 \end{bmatrix} - \begin{bmatrix} 8 & 1 \\ 3 & 4 \end{bmatrix} = \begin{bmatrix} 6-3 & 2-1 \\ 7-3 & -5-4 \end{bmatrix} = \begin{bmatrix} -2 & 1 \\ 4 & -9 \end{bmatrix}$

If A, B and C are the matrices of the same order mxn, then A + B = B + A,

and (A + B) + C = A + (B + C) i.e., the addition of matrices is commutative and Associative respectively.

The sum or difference of two matrices of different order is not defined.

Product of Matrices

Two matrices A and B are said to be conformable for the product AB if the number of columns of A is equal to the number of rows of B.

Then the product matrix AB has the same number of rows as A and the same number of columns as B.

Thus the product of the matrices A_{mxp} and B_{pxn} is the matrix $(AB)_{mxn}$. The elements of AB are determined as follows:

The element C_{ij} in the ith row and jth column of $(AB)_{mxn}$ is found by $c_{ij} = a_{ij}b_{1j} + a_{i2}b_{2j} + a_{i3}b_{3j} + \dots\dots\dots + a_{in}b_{nj}$.

For example, consider the matrices,

$$A_{2x2}\begin{bmatrix} a_{11} & a_{12} \\ a_{21} & a_{22} \end{bmatrix} \text{and } B_{2x2} = \begin{bmatrix} b_{11} & b_{12} \\ b_{21} & b_{22} \end{bmatrix}$$

Since the number of columns of A is equal to the number of rows of B, the product AB is defined and is given as:

$$AB = \begin{bmatrix} a_{11}b_{11} + a_{12}b_{21} & a_{11}b_{12} + a_{12}b_{22} \\ a_{21}b_{11} + a_{22}b_{21} & a_{21} + b_{12} + a_{22}b_{22} \end{bmatrix}$$

Thus c_{11} is obtained by multiplying the elements of the first row of A i.e., a_{11}, a_{12} by the corresponding elements of the first column of B i.e., b_{11}, b_{21} and adding the product.

Similarly, c_{12} is obtained by multiplying the elements of the first row of A i.e., a_{11}, a_{12} by the corresponding elements of the second column of B i.e., b_{12}, b_{22} and adding the product. Similarly for c_{21}, c_{22}.

- Multiplication of matrices is not commutative i.e., AB ≠ BA in general.
- For matrices A and B if AB = BA then A and B commute to each other.
- A matrix A can be multiplied by itself if and only if it is a square matrix. The product A.A in such cases is written as A^2.

Similarly we may define higher powers of a square matrix i.e., A. $A^2 = A^3$, A^2. $A^2 = A^4$

In the product AB, A is said to be pre multiple of B and B is said to be post multiple of A.

Example: If $A = \begin{bmatrix} 1 & 2 \\ -1 & 3 \end{bmatrix}$ and $B = \begin{bmatrix} 2 & 1 \\ 1 & 1 \end{bmatrix}$ Find AB and BA.

Solution:

$$AB = \begin{bmatrix} 1 & 2 \\ -1 & 3 \end{bmatrix}\begin{bmatrix} 2 & 1 \\ 1 & 1 \end{bmatrix} = \begin{bmatrix} 2+2 & 1+2 \\ -2+3 & -1+3 \end{bmatrix}$$

$$= \begin{bmatrix} 4 & 3 \\ 1 & 2 \end{bmatrix}$$

$$BA = \begin{bmatrix} 2 & 1 \\ 1 & 1 \end{bmatrix}\begin{bmatrix} 1 & 2 \\ -1 & 3 \end{bmatrix} = \begin{bmatrix} 2-2 & 4+3 \\ 1-1 & 2+3 \end{bmatrix}$$

$$= \begin{bmatrix} 1 & 7 \\ 0 & 5 \end{bmatrix}$$

This example shows very clearly that multiplication of matrices in general, is not commutative i.e., AB ≠ BA.

Example: If $A = \begin{bmatrix} 3 & 1 & 2 \\ 1 & 0 & 1 \end{bmatrix}$ and $B = \begin{bmatrix} 1 & -1 \\ 2 & 1 \\ 3 & 1 \end{bmatrix}$, find AB.

Solution:

Since A is a (2 x 3) matrix and B is a (3 x 2) matrix, they are conformable for multiplication. We have,

$$AB = \begin{bmatrix} 3 & 1 & 2 \\ 1 & 0 & 1 \end{bmatrix}\begin{bmatrix} 1 & -1 \\ 2 & 1 \\ 3 & 1 \end{bmatrix} = \begin{bmatrix} 3+2+6 & -3+1+2 \\ 1+0+3 & -1+0+1 \end{bmatrix}$$

$$= \begin{bmatrix} 11 & 0 \\ 4 & 0 \end{bmatrix}$$

If A, B and C are the matrices of order (m x p), (p x q) and (q x n) respectively, then:

- (AB)C = A(BC) i.e., Associative law holds.
- C(A+B) = CA + CB.

and (A + B)C = AC + BC i.e distributive laws holds.

If a matrix A and identity matrix I are conformable for multiplication, then I has the property that.

AI = IA = A i.e, I is the identity matrix for multiplication.

Special Matrices

Transpose of a Matrix

If $A = [a_{ij}]$ is mxn matrix, then the matrix of order n x m obtained by interchanging the rows and columns of A is called the transpose of A. It is denoted A^t or A'.

Example: if $A = \begin{bmatrix} 1 & 2 & 3 \\ 4 & 5 & 6 \\ 7 & 8 & 9 \end{bmatrix}$, then $A^t = \begin{bmatrix} 1 & 4 & 7 \\ 2 & 5 & 8 \\ 3 & 6 & 9 \end{bmatrix}$

Symmetric Matrix

A square matrix A is called symmetric if $A = A^t$ for example if,

$$A = \begin{bmatrix} a & b & c \\ b & d & e \\ c & e & f \end{bmatrix}, \text{ then } A^t \begin{bmatrix} a & b & c \\ b & d & e \\ c & e & f \end{bmatrix} = A$$

Thus A is symmetric.

Skew Symmetric

A square matrix A is called skew symmetric if $A = A^t$,

for example if $B = \begin{bmatrix} 0 & -4 & 1 \\ 4 & 0 & -3 \\ -1 & 3 & 0 \end{bmatrix}$, then:

$$B^t = \begin{bmatrix} 0 & 4 & -1 \\ -4 & 0 & 3 \\ 1 & -3 & 0 \end{bmatrix} = (-1)\begin{bmatrix} 0 & -4 & 1 \\ 4 & 0 & -3 \\ -1 & 3 & 0 \end{bmatrix}$$

$$B^t = -B$$

Thus matrix B is skew symmetric.

Singular and Non-Singular Matrices

A square matrix A is called singular if $|A| = 0$ and is non-singular if $|A| \neq 0$, for example if:

$A = \begin{bmatrix} 3 & 2 \\ 9 & 6 \end{bmatrix}$, then $|A| = 0$, Hence A is singular,

and if $A = \begin{bmatrix} 3 & 1 & 6 \\ -1 & 3 & 2 \\ 1 & 0 & 0 \end{bmatrix}$, then $|A| \neq 0$.

Hence A is non-singular.

Example: Find k If $A = \begin{bmatrix} k-2 & 1 \\ 5 & k+2 \end{bmatrix}$ is singular.

Solution: Since A is singular, so $\begin{vmatrix} k-2 & 1 \\ 5 & k+2 \end{vmatrix} = 0.$

$$(k-2)(k+2)-5=0$$
$$k^2-4-5=0$$
$$k^2-9=0 \Rightarrow K=\pm 3.$$

Adjoint and Inverse of a Matrix

Let the determinant of a square matrix A be $|A|$

$$\text{If } A=\begin{bmatrix} a_{11} & a_{12} & a_{13} \\ a_{21} & a_{22} & a_{23} \\ a_{31} & a_{32} & a_{33} \end{bmatrix} \text{ Then } |A|=\begin{vmatrix} a_{11} & a_{12} & a_{13} \\ a_{21} & a_{22} & a_{23} \\ a_{31} & a_{32} & a_{33} \end{vmatrix}$$

The matrix formed by the cofactors of the elements is $\begin{bmatrix} A_{11} & A_{12} & A_{13} \\ A_{21} & A_{22} & A_{23} \\ A_{31} & A_{32} & A_{33} \end{bmatrix}$

Where $A_{11}=(-1)^{1+1}\begin{vmatrix} a_{22} & a_{23} \\ a_{32} & a_{33} \end{vmatrix}=a_{22}a_{33}-a_{23}.a_{32}$

$$A_{12}=(-1)^{1+2}\begin{vmatrix} a_{21} & a_{23} \\ a_{31} & a_{33} \end{vmatrix}=-a_{21}.a_{33}+a_{23}.a_{31};A_{13}=(-1)^{1+3}\begin{vmatrix} a_{21} & a_{22} \\ a_{31} & a_{32} \end{vmatrix}=a_{21}\,a_{32}-a_{22}\,a_{31};$$

$$A_{21}=(-1)^{2+1}\begin{vmatrix} a_{12} & a_{13} \\ a_{32} & a_{33} \end{vmatrix}=-a_{12}.a_{33}+a_{13}.a_{32};A_{22}=(-1)^{2+2}\begin{vmatrix} a_{11} & a_{13} \\ a_{31} & a_{33} \end{vmatrix}=a_{11}\,a_{33}-a_{13}\,a_{31};$$

$$A_{23}=(-1)^{2+3}\begin{vmatrix} a_{11} & a_{12} \\ a_{31} & a_{32} \end{vmatrix}=-a_{11}.a_{32}+a_{12}.a_{31};A_{31}=(-1)^{3+1}\begin{vmatrix} a_{12} & a_{13} \\ a_{22} & a_{23} \end{vmatrix}=a_{21}\,a_{23}-a_{13}\,a_{22};$$

$$A_{32}=(-1)^{3+2}\begin{vmatrix} a_{11} & a_{13} \\ a_{21} & a_{23} \end{vmatrix}=-a_{11}.a_{23}+a_{13}.a_{21};A_{33}=(-1)^{3+3}\begin{vmatrix} a_{11} & a_{12} \\ a_{21} & a_{22} \end{vmatrix}=a_{11}\,a_{22}-a_{12}\,a_{21};$$

Then the transpose of the matrix of co-factors is called the adjoint of the matrix A and is written as adj A.

$$\text{adj } A=\begin{bmatrix} A_{11} & A_{21} & A_{31} \\ A_{12} & A_{22} & A_{32} \\ A_{13} & A_{23} & A_{33} \end{bmatrix}$$

The product of a matrix A and its adjoint is equal to unit matrix multiplied by the determinant A.

Let A be a square matrix, then (Adjoint A). A = A. (Adjoint A) = | A |. I

$$\text{Let } A=\begin{bmatrix} a_{11} & a_{12} & a_{13} \\ a_{21} & a_{22} & a_{23} \\ a_{31} & a_{32} & a_{33} \end{bmatrix} \text{ and adj } A=\begin{bmatrix} A_{11} & A_{21} & A_{31} \\ A_{12} & A_{22} & A_{32} \\ A_{13} & A_{23} & A_{33} \end{bmatrix}$$

$$A.\ (\text{adj. } A) = \begin{bmatrix} a_{11} & a_{12} & a_{13} \\ a_{21} & a_{22} & a_{23} \\ a_{31} & a_{32} & a_{33} \end{bmatrix} \times \begin{bmatrix} A_{11} & A_{21} & A_{31} \\ A_{12} & A_{22} & A_{32} \\ A_{13} & A_{23} & A_{33} \end{bmatrix}$$

$$\begin{bmatrix} a_{11}A_{11} + a_{12}A_{12} + a_{13}A_{13} & a_{11}A_{21} + a_{12}A_{22} + a_{13}A_{23} & a_{11}A_{31} + a_{12}A_{32} + a_{13}A_{33} \\ a_{21}A_{11} + a_{22}A_{12} + a_{23}A_{13} & a_{21}A_{21} + a_{22}A_{22} + a_{23}A_{23} & a_{21}A_{31} + a_{22}A_{32} + a_{23}A_{33} \\ a_{31}A_{11} + a_{32}A_{12} + a_{33}A_{13} & a_{31}A_{21} + a_{32}A_{22} + a_{33}A_{23} & a_{31}A_{31} + a_{32}A_{32} + a_{33}A_{33} \end{bmatrix}$$

$$= \begin{bmatrix} |A| & 0 & 0 \\ 0 & |A| & 0 \\ 0 & 0 & |A| \end{bmatrix} = |A| \begin{bmatrix} 1 & 0 & 0 \\ 0 & 1 & 0 \\ 0 & 0 & 1 \end{bmatrix} = |A|\ I.$$

How to Find the Adjoint of a Matrix and Its Inverse?

Example on How to Find the Adjoint of a Matrix

Example: If $A^T = -A$ then the elements on the diagonal of the matrix are equal to:

- 1
- -1
- 0
- None of these

Solution:

(c) $A^T = -A$; A is skew-symmetric matrix; diagonal elements of A are zeros. so option (c) is the answer.

Example: If A and B are two skew-symmetric matrices of order n, then,

- AB is a skew-symmetric matrix.
- AB is a symmetric matrix.
- AB is a symmetric matrix if A and B commute.
- None of these.

Solution:

- We are given A' = -A and B' = -B;

  Now, (AB)' = B'A' = (-B) (-A) = BA = AB, if A and B commute.

Example: Let A and B be two matrices such that AB' + BA' = 0. If A is skew symmetric, then BA

- Symmetric.
- Skew symmetric.

- Invertible.
- None of these.

Solution: We have, (BA)' = A'B' = -AB' [A is skew symmetric]; = BA' = B(-A) = -BA

BA is skew symmetric.

Example: Let A = $\begin{bmatrix} 1 & 2 & 3 \\ 1 & 3 & 4 \\ 1 & 4 & 3 \end{bmatrix}$,

then adj A is given by.

Solution:

Co-factors of the elements of any matrix are obtained by eliminating all the elements of the same row and column and calculating the determinant of the remaining elements.

$$A_{11} = \begin{vmatrix} 3 & 4 \\ 4 & 3 \end{vmatrix} = 3\times3-4\times4 = -7$$

$$A_{12} = -\begin{vmatrix} 1 & 4 \\ 1 & 3 \end{vmatrix} = 1,\ A_{13} = \begin{vmatrix} 1 & 3 \\ 1 & 4 \end{vmatrix} = 1;\ A_{21} = -\begin{vmatrix} 2 & 3 \\ 4 & 3 \end{vmatrix} = 6, A_{22} = \begin{vmatrix} 1 & 3 \\ 1 & 3 \end{vmatrix} = 0$$

$$A_{23} = -\begin{vmatrix} 1 & 2 \\ 1 & 4 \end{vmatrix} = -2,\ A_{31} = \begin{vmatrix} 2 & 3 \\ 3 & 4 \end{vmatrix} = -1;\ A_{32} = -\begin{vmatrix} 1 & 3 \\ 1 & 4 \end{vmatrix} = -1,\ A_{33} = \begin{vmatrix} 1 & 2 \\ 1 & 3 \end{vmatrix} = 1$$

adj A = transpose of cofactor matrix.

$$\therefore \text{Adj A} = \begin{vmatrix} -7 & 6 & -1 \\ 1 & 0 & -1 \\ 1 & -2 & 1 \end{vmatrix}$$

Example: Which of the following statements are false:

- If | A | = 0, then | adj A | = 0.
- Adjoint of a diagonal matrix of order 3 × 3 is a diagonal matrix.
- Product of two upper triangular matrices is an upper triangular matrix.
- adj (AB) = adj (A) adj (B).

Solution: (d) We have, adj (AB) = adj (B) adj (A) and not adj (AB) = adj (A) adj (B).

Inverse of a Matrix

If A and B are two square matrices of the same order, such that AB = BA = I (I = unit matrix).

Then B is called the inverse of A, i.e. B = A^{-1} and A is the inverse of B. Condition for a square matrix

A to possess an inverse is that the matrix A is non-singular, i.e., $|A| \neq 0$. If A is a square matrix and B is its inverse then AB = I. Taking determinant of both sides $|AB| = |I|$ or $|A||B| = I$. From this relation it is clear that $|A| \neq 0$, i.e. the matrix A is non-singular.

How to find the Inverse of a Matrix by using the Adjoint Matrix?

We know that, $A.(\text{Adj } A) = |A| I$ or $\frac{A.(\text{Adj } A)}{|A|} = I(\text{Provided } |A| \neq 0)$

And $A.A^{-1} = I; \; A^{-1} = \frac{1}{|A|}(\text{Adj}.A)$.

Properties of Inverse and Adjoint of a Matrix

- Property 1: For a square matrix A of order n, A adj(A) = adj(A) A = |A|I, where I is the identitiy matrix of order n.
- Property 2: A square matrix A is invertible if and only if A is a non-singular matrix.

Problems on Finding the Inverse of a Matrix

Examples: Let $A = \begin{bmatrix} 1 & 0 & -1 \\ 3 & 4 & 5 \\ 0 & -6 & -7 \end{bmatrix}$. What is inverse of A?

Solution:

By using the formula $A^{-1} = \frac{\text{adj } A}{|A|}$, we can obtain the value of A^{-1}

We have $A_{11} = \begin{bmatrix} 4 & 5 \\ -6 & -7 \end{bmatrix} = 2 \; A_{12} = -\begin{bmatrix} 3 & 5 \\ 0 & -7 \end{bmatrix} = 21$

Similarly $A_{13} = -18, A_{31} = 4, A_{32} = -8, A_{33} = 4, A_{21} = +6, A_{22} = -7, A_{23} = 6$

cofactor matrix of A= $\begin{bmatrix} 2 & 21 & -18 \\ 6 & -7 & 6 \\ 4 & -8 & 4 \end{bmatrix}$

adj A = transpose of cofactor matrix

$$\text{adj } A = \begin{bmatrix} 2 & 6 & 4 \\ 21 & -7 & -8 \\ -18 & 6 & 4 \end{bmatrix}$$

Also $|A| = \begin{vmatrix} 1 & 0 & -1 \\ 3 & 4 & 5 \\ 0 & -6 & -7 \end{vmatrix} = \{4\times(-7)-(-6)\times5-3\times(-6)\}$

$$= -28+30+18$$
$$= 20$$

$$A^{-1} = \frac{\text{adj A}}{|A|} = \frac{1}{20}\begin{bmatrix} 2 & 6 & 4 \\ 21 & -7 & -8 \\ -18 & 6 & 4 \end{bmatrix}$$

Examples: If the product of a matrix A and $\begin{bmatrix} 1 & 1 \\ 2 & 0 \end{bmatrix}$ is the matrix $\begin{bmatrix} 3 & 2 \\ 1 & 1 \end{bmatrix}$, then A^{-1} is given by:

- $\begin{bmatrix} 0 & -1 \\ 2 & -4 \end{bmatrix}$ (b) $\begin{bmatrix} 0 & -1 \\ -2 & -4 \end{bmatrix}$ (c) $\begin{bmatrix} 0 & 1 \\ 2 & -4 \end{bmatrix}$.
- None of these.

Solution:

We know if AB = C, then $B^{-1}A^{-1} = C^{-1} \Rightarrow A^{-1} = BC^{-1}$ by using this formula we will get value of A^{-1}in the above problem.

Here, $A\begin{bmatrix} 1 & 1 \\ 2 & 0 \end{bmatrix} = \begin{bmatrix} 3 & 2 \\ 1 & 1 \end{bmatrix} \Rightarrow A^{-1} = \begin{bmatrix} 1 & 1 \\ 2 & 0 \end{bmatrix}\begin{bmatrix} 3 & 2 \\ 1 & 1 \end{bmatrix}^{-1} = \begin{bmatrix} 1 & 1 \\ 2 & 0 \end{bmatrix}\begin{bmatrix} 1 & -2 \\ -1 & 3 \end{bmatrix} = \begin{bmatrix} 0 & 1 \\ 2 & -4 \end{bmatrix}$

Examples:

Let $A = \begin{bmatrix} 2 & 1 & -1 \\ 0 & 1 & 0 \\ 1 & 3 & -1 \end{bmatrix}$ and $B = \begin{bmatrix} 1 & 2 & 5 \\ 2 & 3 & 1 \\ -1 & 1 & 1 \end{bmatrix}$. Prove that $(AB)^{-1} = B^{-1}A^{-1}$.

Solution:

By obtaining | AB | and adj AB we can obtain $(AB)^{-1}$ by using the formula $(AB)^{-1} = \frac{\text{adj AB}}{|AB|}$.

Similarly we can also obtain the values of B^{-1} and A^{-1} Then by multiplying B^{-1} and A^{-1} we can prove the given problem.

Here, $AB = \begin{bmatrix} 2 & 1 & -1 \\ 0 & 1 & 0 \\ 1 & 3 & -1 \end{bmatrix}\begin{bmatrix} 1 & 2 & 5 \\ 2 & 3 & 1 \\ -1 & 1 & 1 \end{bmatrix} = \begin{bmatrix} 2+2+1 & 4+3-1 & 10+1-1 \\ 0+2+0 & 0+3+0 & 0+1+0 \\ 1+6+1 & 2+9-1 & 5+3-1 \end{bmatrix} = \begin{bmatrix} 5 & 6 & 10 \\ 2 & 3 & 1 \\ 8 & 10 & 7 \end{bmatrix}$

Now, $|AB| = \begin{vmatrix} 5 & 6 & 10 \\ 2 & 3 & 1 \\ 8 & 10 & 7 \end{vmatrix} = 5(21-10)-6(14-8)+10(20-24) = 55-36-40 = -21.$

The matrix of cofactors of | AB | is equal to:

$$\begin{bmatrix} 3(7)-1(10) & -\{2(7)-8(1)\} & 2(10)-3(8) \\ -\{6(7)-10(10)\} & 5(7)-8(10) & -\{5(10)-6(8)\} \\ 6(1)-10(3) & -\{5(1)-2(10)\} & 5(3)-6(2) \end{bmatrix} = \begin{bmatrix} 11 & -6 & -4 \\ 58 & -45 & -2 \\ -24 & 15 & 3 \end{bmatrix}$$

$$\text{adj AB} = \begin{bmatrix} 11 & 58 & -24 \\ -6 & -45 & 15 \\ -4 & -2 & 3 \end{bmatrix} \text{ So, } (AB)^{-1} = \frac{\text{adj AB}}{|AB|} = \frac{-1}{21}\begin{bmatrix} 11 & 58 & -24 \\ -6 & -45 & 15 \\ -4 & -2 & 3 \end{bmatrix}$$

Next, $|B| = \begin{vmatrix} 1 & 2 & 5 \\ 2 & 3 & 1 \\ -1 & 1 & 1 \end{vmatrix} = 1(3-1)-2(2+1)+5(2+3) = 21$

Cofactor matrix of B = $\begin{bmatrix} 2 & -3 & 5 \\ 3 & 6 & -3 \\ -13 & 9 & -1 \end{bmatrix}$

$$\text{adj B} = \begin{bmatrix} 2 & 3 & -13 \\ -3 & 6 & 9 \\ 5 & -3 & -1 \end{bmatrix}$$

$$\therefore B^{-1} = \frac{\text{adj B}}{|B|} = \frac{1}{21}\begin{bmatrix} 2 & 3 & -13 \\ -3 & 6 & 9 \\ 5 & -3 & -1 \end{bmatrix}; \ |A| = \begin{vmatrix} 2 & 1 & -1 \\ 0 & 1 & 0 \\ 1 & 3 & -1 \end{vmatrix} = 1(-2+1) = -1$$

Similarly cofactor matrix of A = $\begin{bmatrix} -1 & 0 & -1 \\ -2 & -1 & -5 \\ 1 & 0 & 2 \end{bmatrix}$

$$A^{-1} = \frac{\text{adj A}}{|A|} = \frac{1}{-1}\begin{bmatrix} -1 & -2 & 1 \\ 0 & -1 & 0 \\ -1 & -5 & 2 \end{bmatrix}$$

$$\therefore B^{-1}A^{-1} = -\frac{1}{21}\begin{bmatrix} 2 & 3 & -13 \\ -3 & 6 & 9 \\ 5 & -3 & -1 \end{bmatrix}\begin{bmatrix} -1 & -2 & 1 \\ 0 & -1 & 0 \\ -1 & -5 & 2 \end{bmatrix}$$

$$= -\frac{1}{21}\begin{bmatrix} 11 & 58 & -24 \\ -6 & -45 & 15 \\ -4 & -2 & 3 \end{bmatrix} \text{ Thus } (AB)^{-1} = B^{-1}A^{-1}$$

Examples: If $A = \begin{bmatrix} 0 & 2y & z \\ x & y & -z \\ x & -y & z \end{bmatrix}$ satisfies $A' = A^{-1}$,

Then:

- $x = \pm 1\sqrt{6}, y = \pm 1/\sqrt{6}, z = \pm 1/\sqrt{3}$.
- $x = \pm 1\sqrt{2}, y = \pm 1/\sqrt{6}, z = \pm 1/\sqrt{3}$.
- $x = \pm 1/\sqrt{6}, y = \pm 1/\sqrt{2}, z = \pm 1/\sqrt{3}$.
- $x = \pm 1/\sqrt{2}, y = \pm 1/3, z = \pm/\sqrt{2}$.

Solution:

Given that $A' = A^{-1}$ and we know that $AA^{-1} = I$ and therefore $AA' = I$. Using the multiplication method we can obtain values of x, y and z.

$$A' = A^{-1} \Leftrightarrow AA' = 1$$

Now,

$$AA' = \begin{bmatrix} 0 & 2y & z \\ x & y & -z \\ x & -y & z \end{bmatrix}\begin{bmatrix} 0 & x & x \\ 2y & y & -y \\ z & -z & z \end{bmatrix} = \begin{bmatrix} 4y^2 + z^2 & 2y^2 - z^2 & -2y^2 + z^2 \\ 2y^2 - z^2 & x^2 + y^2 + z^2 & x^2 - y^2 - z^2 \\ -2y^2 + z^2 & x^2 - y^2 - z^2 & x^2 + y^2 + z^2 \end{bmatrix}$$

Thus, $AA' = I \Rightarrow 4y^2 + z^2 = 1,\ 2y^2 - z^2 = 0,\ x^2 + y^2 + z^2 = 1,\ x^2 - y^2 - z^2 = 0$

$$x = \pm 1/\sqrt{2},\ y = \pm 1/\sqrt{6},\ z = \pm 1/\sqrt{3}$$

Examples: If $A = \begin{bmatrix} 0 & 1 & 2 \\ 1 & 2 & 3 \\ 3 & x & 1 \end{bmatrix}$ and $A^{-1} = \begin{bmatrix} 1/2 & -1/2 & 1/2 \\ -4 & 3 & y \\ 5/2 & -3/2 & 1/2 \end{bmatrix}$,

Then

$(a) x=1, y=-1$

$(b) x=-1, y=1$

$(c) x=2, y=-1/2$

$(d) x=1/2, y=\frac{1}{2}$

Solution:

We know $AA^{-1}=I$, hence by solving it we can obtain the values of x and y.

We have,

$$\begin{bmatrix} 1 & 0 & 0 \\ 0 & 1 & 0 \\ 0 & 0 & 1 \end{bmatrix} = AA^{-1} = \begin{bmatrix} 0 & 1 & 2 \\ 1 & 2 & 3 \\ 3 & x & 1 \end{bmatrix} \begin{bmatrix} 1/2 & -1/2 & 1/2 \\ -4 & 3 & y \\ 5/2 & -3/2 & 1/2 \end{bmatrix} =$$

$$\begin{bmatrix} 1 & 0 & y+1 \\ 0 & 1 & 2(y+1) \\ 4(1-x) & 3(x-1) & 2+xy \end{bmatrix}$$

$$\Rightarrow 1-x=0,\ x-1=0;\ y+1=0,\ y+1=0,\ 2+xy=1$$

$$\therefore x=1,\ y=-1$$

Determinants

The determinant has a variety of applications. The value of the determinant of a square matrix A can be used to determine whether A is invertible or noninvertible. An explicit formula for A^{-1} exists that involves the determinant of A. Some systems of linear equations have solutions that can be expressed in terms of determinants.

The determinant of the 2 × 2 matrix $\begin{bmatrix} a_{11} & a_{12} \\ a_{21} & a_{22} \end{bmatrix}$ was defined to be the number $a_{11}a_{22}-a_{12}a_{21}$ and that the notation det (A) or |A| was used to represent the determinant of A. For any given n × n matrix $A=[a_{ij}]_{n\times n}$, the notation A_{ij} will be used to denote the $(n-1)\times(n-1)$ submatrix obtained from A by deleting the ith row and the jth column of A. The determinant of any size square matrix $A=\lfloor a_{ij}\rfloor_{n\times n}$ is defined recursively as follows.

Definition of the Determinant: Let $A=[a_{ij}]_{n\times n}$ be an n × n matrix:

- If n = 1, that is $A=[a_{11}]$, then we define det $(A)=a_{11}$.

- If n > 1, we define $\det(A)=\sum_{k=1}^{n}(-1)^{1+k}a_{1k}\det(A_{1k})$.

Example:

If $A=[5]$, then by part (1) of the definition of the determinant, det (A) = 5.

If $A=\begin{bmatrix}2&3\\4&5\end{bmatrix}$, then by parts (2) and (1), $\det(A)=(-1)^{1+1}(2)\det[5]+(-1)^{1+2}(3)\det[4]$

$=(1)(2)(5)+(-1)(3)(4)=10-12=-2$

If $A=\begin{bmatrix}2&3&4\\5&6&7\\8&9&1\end{bmatrix}$, then using parts (2) and (1), we calculate the det (A) as follows:

$$\det(A)=(-1)^{1+1}(2)\det\begin{bmatrix}6&7\\9&1\end{bmatrix}+(-1)^{1+2}(3)\det\begin{bmatrix}5&7\\8&1\end{bmatrix}+(-1)^{1+3}(4)\det\begin{bmatrix}5&6\\8&9\end{bmatrix}$$
$$=(1)(2)(-57)+(-1)(3)(-51)+(1)(4)(-3)=-114+153-12=27.$$

Cofactor

If A is a square matrix, the ij[th] cofactor of A is defined to be $(-1)^{i+j}\det(A_{ij})$. The notation C_{ij} will sometimes be used to denote the ij[th] cofactor of A.

Example: Let $A=\begin{bmatrix}1&2&3\\4&5&6\\7&8&9\end{bmatrix}$. Then $C_{11}=(-1)^{1+1}\det\begin{bmatrix}5&6\\8&9\end{bmatrix}=(1)(45-48)=-3$,

$$C_{12}=(-1)^{1+2}\det\begin{bmatrix}4&6\\7&9\end{bmatrix}=(-1)(36-42)=6$$

and,

$$C_{23}=(-1)^{2+3}\det\begin{bmatrix}1&2\\7&8\end{bmatrix}=(-1)(8-14)=6.$$

In the definition of the determinant, part (2) consists of multiplying each first row entry of A by its cofactor and then summing these products. For this reason it is called a first row cofactor expansion.

Example: Let $A=\begin{bmatrix}2&3&4&5\\1&3&2&4\\5&3&4&1\\4&2&3&5\end{bmatrix}$. Use a first row cofactor expansion to evaluate det(A).

Solution: $\det(A) =$

$$(-1)^{1+1}(2)\det\begin{bmatrix}3&2&4\\3&4&1\\2&3&5\end{bmatrix}+(-1)^{1+2}(3)\det\begin{bmatrix}1&2&4\\5&4&1\\4&3&5\end{bmatrix}+(-1)^{1+3}(4)\det\begin{bmatrix}1&3&4\\5&3&1\\4&2&5\end{bmatrix}$$

$$+(-1)^{1+4}(5)\det\begin{bmatrix}1&3&2\\5&3&4\\4&2&3\end{bmatrix}$$

$$=(1)(2)\left\{(-1)^{1+1}(3)\det\begin{bmatrix}4&1\\3&5\end{bmatrix}+(-1)^{1+2}(2)\det\begin{bmatrix}3&1\\2&5\end{bmatrix}+(-1)^{1+3}(4)\det\begin{bmatrix}3&4\\2&3\end{bmatrix}\right\}$$

$$+(-1)(3)\left\{(-1)^{1+1}(1)\det\begin{bmatrix}4&1\\3&5\end{bmatrix}+(-1)^{1+2}(2)\det\begin{bmatrix}5&1\\4&5\end{bmatrix}+(-1)^{1+3}(4)\det\begin{bmatrix}5&4\\4&3\end{bmatrix}\right\}$$

$$+(1)(4)\left\{(-1)^{1+1}(1)\det\begin{bmatrix}3&1\\1&5\end{bmatrix}+(-1)^{1+2}(3)\det\begin{bmatrix}5&1\\4&5\end{bmatrix}+(-1)^{1+3}(4)\det\begin{bmatrix}5&3\\4&2\end{bmatrix}\right\}$$

$$+(-1)(5)\left\{(-1)^{1+1}(1)\det\begin{bmatrix}3&4\\2&3\end{bmatrix}+(-1)^{1+2}(3)\det\begin{bmatrix}5&4\\4&3\end{bmatrix}+(-1)^{1+3}(2)\det\begin{bmatrix}5&3\\4&2\end{bmatrix}\right\}$$

$$=(1)(2)\{(1)(3)(20-3)+(-1)(2)(15-2)+(1)(4)(9-8)\}$$

$$+(-1)(3)\{(1)(1)(20-3)+(-1)(2)(25-4)+(1)(4)(15-16)\}$$

$$+(1)(4)\{(1)(1)(15-2)+(-1)(3)(25-4)+(1)(4)(10-12)\}$$

$$+(-1)(5)\{(1)(1)(9-8)+(-1)(3)(15-16)+(1)(2)(10-12)\}$$

$$=(1)(2)\{51-26+4\}+(-1)(3)\{17-42-4\}+(1)(4)\{13-63-8\}+(-1)(5)\{1+3-4\}$$

$$=58+87-232+0=-87$$

Although the definition of the determinant uses a first row cofactor expansion, the determinant of A may be calculated by taking any row (or column) and multiplying the entries of that row (or column) by their cofactors and summing the products. This result is given in the next theorem whose proof is omitted.

Theorem: Let A be a square n × n matrix, then:

$$\det(A)=\sum_{k=1}^{n}(-1)^{i+k}a_{ik}\det(A_{ik})=\sum_{k=1}^{n}(-1)^{k+j}a_{kj}\det(A_{kj}).$$

↑ i^{th} row cofactor expansion

↑ j^{th} column cofactor expansion

Example: Let $A = \begin{bmatrix} 1 & 3 & 4 \\ 2 & 0 & 1 \\ 3 & 1 & 0 \end{bmatrix}$. Evaluate det (A) by

- a second row cofactor expansion.
- a third column cofactor expansion.

Solution:

$$(a)\det(A) = (-1)^{2+1}(2)\det\begin{bmatrix} 3 & 4 \\ 1 & 0 \end{bmatrix} + (-1)^{2+2}(0)\det\begin{bmatrix} 1 & 4 \\ 3 & 0 \end{bmatrix} + (-1)^{2+3}(1)\det\begin{bmatrix} 1 & 3 \\ 3 & 1 \end{bmatrix}$$
$$= (-1)(2)(-4) + (1)(0)(-12) + (-1)(1)(-8) = 8 + 0 + 8 = 16.$$

$$(a)\det(A) = (-1)^{1+3}(4)\det\begin{bmatrix} 2 & 0 \\ 3 & 1 \end{bmatrix} + (-1)^{2+3}(1)\det\begin{bmatrix} 1 & 3 \\ 3 & 1 \end{bmatrix} + (-1)^{3+3}(0)\det\begin{bmatrix} 1 & 3 \\ 2 & 0 \end{bmatrix}$$
$$= (1)(4)(2) + (-1)(1)(-8) + (1)(0)(-6) = 8 + 8 + 0 = 16.$$

Theorem: If A is a square matrix containing a row (or column) of zeros, then det(A) = 0.

Proof: Use a cofactor expansion along the row (or column) of zeros.

Theorem: If A is an n × n matrix with two identical rows (or columns), then det (A) = 0.

Proof: The theorem is certainly true for n = 2 since $\det\begin{bmatrix} a_{11} & a_{12} \\ a_{11} & a_{12} \end{bmatrix} = a_{11}a_{12} - a_{11}a_{12} = 0.$

If n = 3, use a cofactor expansion along the row different from the two identical rows. Let this row be the kth row. Using a cofactor expansion along this row gives $\det(A) = (-1)^{k+1}(a_{k1})\det(A_{k1}) + (-1)^{k+2}(a_{k2})\det(A_{k2}) + (-1)^{k+3}(a_{k3})\det(A_{k3})$. But each of the submatrices A_{k1}, A_{k2} and A_{k3} has two identical rows so their determinants are 0, hence det (A) = 0 for any 3 × 3 matrix.

If n > 3 proceed as above writing det (A) as a sum of products involving submatrices with two identical rows whose determinants are 0.

Triangular and Diagonal Matrices

A square matrix is said to be an upper triangular matrix if all the entries below the main diagonal are zero. A square matrix is said to be a lower triangular matrix if all the entries above the main diagonal are zero. A square matrix is said to be a diagonal matrix if all entries not on the main diagonal are zero. A diagonal matrix is both upper triangular and lower triangular.

Example:

$\begin{bmatrix} 1 & 2 & 3 \\ 0 & 4 & 5 \\ 0 & 0 & 6 \end{bmatrix}$ is an upper triangular matrix.

$\begin{bmatrix} 1 & 0 & 0 \\ 2 & 3 & 0 \\ 4 & 5 & 6 \end{bmatrix}$ is a lower triangular matrix.

$\begin{bmatrix} 1 & 0 & 0 \\ 0 & 2 & 0 \\ 0 & 0 & 3 \end{bmatrix}$ is a diagonal matrix. It is both upper and lower triangular.

Theorem: If A is upper triangular, lower triangular or diagonal, $\det(A) = a_{11}a_{22}.......a_{nn}$.

Proof: Suppose A is upper triangular. To evaluate det(A) use a cofactor expansion along the first column. Since there is only one nonzero entry in the first column the expansion gives $\det(A) = (-1)^{1+1} a_{11} \det(A_{11}) = a_{11} \det(A_{11})$. Now A_{11} is upper triangular so proceed as above to use a cofactor expansion along its first column to get $\det(A_{11}) = a_{22} \det(A_{22})$ where A_{22} is A_{11} with its first row and first column deleted. Combining the results gives $\det(A) = a_{11}a_{22} \det(A_{22})$. Continuing in this fashion, we eventually get $\det(A) = a_{11}a_{22}.....a_{nn}$ as required. If A is lower triangular or diagonal, the argument is similar.

Example: $\det\begin{bmatrix} 2 & 3 & 4 \\ 0 & 5 & 6 \\ 0 & 0 & 7 \end{bmatrix} = (2)(5)(7) = 70.$ $\det\begin{bmatrix} 2 & 0 & 0 \\ 0 & 3 & 0 \\ 0 & 0 & 5 \end{bmatrix} = (2)(3)(5) = 30$

Theorem: $\det(I_n) = 1$ for all n.

Proof: Since I_n is a diagonal matrix, $\det(I_n) = (1)(1).......(1) = 1$.

Basket-Weave Method

The following method is an alternative way to evaluate the determinant of a 3 × 3 matrix. This method is only applicable to 3 × 3 matrices and is sometimes called the basket-weave method.

Construct a 3 × 5 array by writing down the entries of the 3 × 3 matrix and then repeating the first two columns. Calculate the products along the six diagonal lines shown in the diagram. The determinant is equal to the sum of products along diagonals labeled 1, 2 and 3 minus the sum of the products along the diagonals labeled 4, 5 and 6.

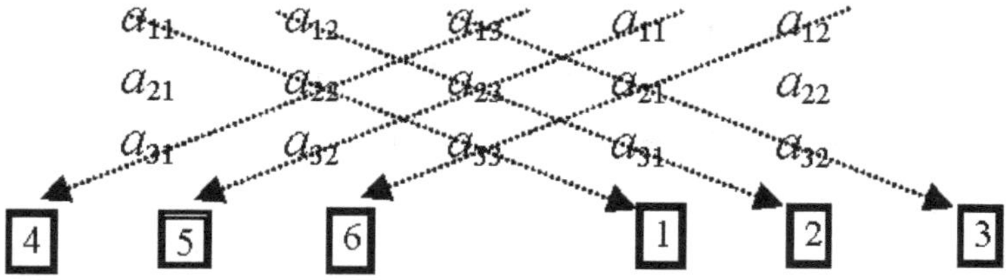

Example: Use the basket-weave method to calculate the determinant of $A = \begin{bmatrix} 1 & 2 & 3 \\ 2 & 1 & 4 \\ 3 & 2 & 5 \end{bmatrix}$.

Solution:

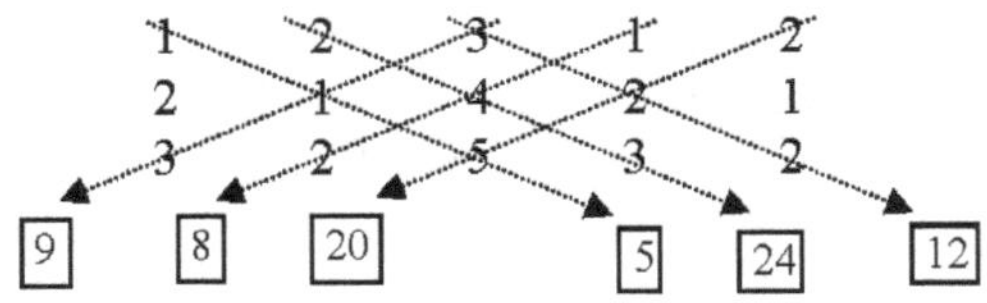

$$\det(A) = (5+24+12) - (9+8+20)$$
$$= (41) - (37) = 4$$

1. Use the definition of the determinant to evaluate the determinant of the given matrix.

$$(a)\begin{bmatrix}1&2&1\\2&3&1\\3&2&2\end{bmatrix} \quad (b)\begin{bmatrix}2&1&3\\4&3&5\\1&0&2\end{bmatrix} \quad (c)\begin{bmatrix}1&1&3\\2&1&4\\5&3&2\end{bmatrix} \quad (d)\begin{bmatrix}1&0&1\\2&0&0\\3&4&5\end{bmatrix}$$

$$(e)\begin{bmatrix}1&0&1&0\\2&0&3&1\\0&2&4&3\\2&3&1&4\end{bmatrix} \quad (f)\begin{bmatrix}3&2&0&1\\5&0&3&2\\0&4&0&2\\3&1&0&2\end{bmatrix} \quad (g)\begin{bmatrix}5&4&3&1\\0&4&1&2\\3&1&2&4\\0&2&3&1\end{bmatrix} \quad (h)\begin{bmatrix}3&0&2&1\\0&2&4&3\\2&3&1&0\\4&1&0&3\end{bmatrix}$$

2. Use the basket-weave method to evaluate the determinants 1(a), 1(b), 1(c) and 1(d).

3. Evaluate the determinants of the following matrices by inspection.

$$(a)\begin{bmatrix}2&3&4\\0&5&6\\0&0&1\end{bmatrix} \quad (b)\begin{bmatrix}3&5&9\\0&6&3\\0&0&2\end{bmatrix} \quad (c)\begin{bmatrix}4&0&0\\5&3&0\\1&3&2\end{bmatrix} \quad (d)\begin{bmatrix}2&0&0\\0&3&0\\0&0&4\end{bmatrix}$$

4. Let $A = \begin{bmatrix}1&2&3\\2&3&5\\6&4&1\end{bmatrix}$. Find the following cofactors of A.

(a) C_{11} (b) C_{12} (c) C_{13} (d) C_{21} (e) C_{22} (f) C_{32}

5. Find all values of k for which $\det\begin{bmatrix}1-k&3\\2&2-k\end{bmatrix} = 0$.

Elementary Row Operations on Determinants

The evaluation of the determinant of an n × n matrix using the definition involves the summation of n! terms, each term being a product of n factors. As n increases, this computation becomes too cumbersome and so another technique has been devised to evaluate the determinant. This technique uses the elementary row operations to reduce the matrix to a triangular form. The effect of each elementary row operation on the value of the determinant is taken into account and then the determinant of the triangular matrix is evaluated by finding the product of the entries on the main diagonal.

Theorem: If A and B are square matrices and B is obtained from A by interchanging two rows (or columns) of A, then det (B) = – det (A).

Proof: Let A and B be n × n matrices. If n = 2, then $A=\begin{bmatrix} a_{11} & a_{12} \\ a_{21} & a_{22} \end{bmatrix}$ and $B=\begin{bmatrix} a_{21} & a_{22} \\ a_{11} & a_{12} \end{bmatrix}$ so $\det(B)=a_{21}a_{12}-a_{11}a_{22}=-(a_{11}a_{22}-a_{12}a_{21})=-\det(A)$.

If n = 3, then we use a cofactor expansion for B along the row that was not interchanged. Let this be row k. Then $\det=(-1)^{k+1}a_{k1}\det(B_{k1})+(-1)a_{k2}\det(B_{k2})+(-1)a_{k3}(B_{k3})$. Each submatrix B_{kj} is the submatrix A_{kj} with its rows interchanged so $\det(B_{kj})=-\det(A_{kj})$.

Hence, $\det(B)=(-1)^{k+1}a_{k1}(-1)\det(A_{k1})+(-1)a_{k2}(-1)\det(A_{k2})+(-1)a_{k3}(-1)\det(A_{k3})$

$$=(-1)\left[(-1)^{k+1}a_{k1}\det(A_{k1})+(-1)a_{k2}\det(A_{k2})+(-1)a_{k3}\det(A_{k3})\right]$$

$=(-1)\det(A)=-\det(A)$ as required.

If n > 3 proceed as above using cofactor expansions along rows that were not interchanged to get the final result. The proof for interchanged columns is similar.

Theorem: Let A and B be n × n matrices with B obtained from A by multiplying all the entries of some row (or column) of A by a scalar k. Then det (B) = k det (A).

Proof: Suppose B is obtained from A by multiplying the entries of the jth row of A by k. Use a cofactor expansion for B along its jth row to evaluate the determinant of B and noting that these cofactors for B are equal to the corresponding cofactors for A we get:

$$\det(B)=(-1)^{j+1}ka_{j1}\det(B_{j1})+(-1)^{j+2}ka_{j2}\det(B_{j2})+\cdots+(-1)^{j+n}ka_{jn}\det(B_{jn})$$

$$=(-1)^{j+1}ka_{j1}\det(A_{j1})+(-1)^{j+2}ka_{j2}\det(A_{j2})+.....+(-1)^{j+n}ka_{jn}\det(B_{jn})$$

$$=k\left[(-1)^{j+1}a_{j1}\det(A_{j1})+(-1)^{j+2}ka_{j2}\det(A_{j2})+.....+(-1)^{j+n}ka_{jn}\det(A_{jn})\right]$$

$$=k\det(A).$$

The proof in the case where B is obtained from A by multiplying a column of A by k is similar.

Example:

5 is a common factor in row 2 of the given matrix.

$$\det\begin{bmatrix} 7 & 3 \\ 5 & 10 \end{bmatrix}=5\det\begin{bmatrix} 7 & 3 \\ 1 & 2 \end{bmatrix}=(5)(14-3)=55$$

Corollary: $\det(kA)=k^n\det(A)$.

Proof: Since all n rows of A are multiplied by the scalar k to get kA, using the above theorem n times gives $\det(kA)=(k)(k)\cdots\cdots(k)\det(A)=k^n\det(A)$.

Example:

(3 is factor in row 1 and row 2)

$$\det\begin{bmatrix}12 & 6\\ 3 & 9\end{bmatrix} = (3)(3)\det\begin{bmatrix}4 & 2\\ 1 & 3\end{bmatrix} = 3^2\det\begin{bmatrix}4 & 2\\ 1 & 3\end{bmatrix} = (9)(12-2) = (9)(10) = 90$$

Example: Let A be a 3×3 matrix and let det(A) = 5. Find det(2A).

Solution:

$$\det(2A) = 2^3\det(A) = (8)(5) = 40$$

Example:

(since matrix has 2 identical rows)

$$\det\begin{bmatrix}1 & 2 & 3\\ 3 & 6 & 9\\ 4 & 5 & 6\end{bmatrix} = 3\det\begin{bmatrix}1 & 2 & 3\\ 1 & 2 & 3\\ 4 & 5 & 6\end{bmatrix} = (3)(0) = 0.$$

Notice that row 2 of the Original matrix is 3 times row 1. This leads to the following corollary.

Corollary: If A is a square matrix that has a row (or column) that is a scalar multiple of another row (or column), then det (A) = 0.

Proof: Suppose the jth row of A is k times the ith row of A. Then $\det(A) = k\det(\hat{A})$ where $\hat{A}$ is the matrix A with the jth row multiplied by 1/k. But $\hat{A}$ has two identical rows, (row i = row j), so $\det(\hat{A}) = 0$. Hence det (A) = k det $(\hat{A}) = k\cdot 0 = 0$.

Theorem: Let A be a square matrix and let B be the matrix obtained from A by adding a multiple of one row (or column) of A to another row (or column) of A. Then det (B) = det (A).

Proof: Suppose B is obtained from A by adding c times row i to row j.

Evaluate det (B) using a cofactor expansion along row j. Then,

$$\det(B) = \sum_{k=1}^{n}(-1)^{j+k} b_{jk}\det(B_{jk}) = \sum_{k=1}^{n}(-1)^{j+k}(ca_{ik}+a_{jk})\det(B_{jk})$$

$$= \sum_{k=1}^{n}(-1)^{j+k}(ca_{ik}+a_{jk})\det(A_{jk}) \quad \text{since } B_{jk} = A_{jk} \text{ for all k.}$$

$$= c\sum_{k=1}^{n}(-1)^{j+k}a_{ik}\det(A_{jk}) + \sum_{k=1}^{n}(-1)^{j+k}a_{jk}\det(A_{jk})$$

$= c\det(\hat{A}) + \det(A)$ where $\hat{A}$ is obtained from A by replacing the jth row of A by its ith row.

$= c\cdot 0 + \det(A)$ $\det(\hat{A}) = 0$ because $\hat{A}$ has two identical rows

= det (A)

Example: Evaluate $\det\begin{bmatrix} 2 & 6 & 9 \\ 1 & 2 & 4 \\ 3 & 6 & 15 \end{bmatrix}$

Solution:

$$\det\begin{bmatrix} 2 & 6 & 9 \\ 1 & 2 & 4 \\ 3 & 6 & 15 \end{bmatrix} = (-1)\det\begin{bmatrix} 1 & 2 & 4 \\ 2 & 6 & 9 \\ 3 & 6 & 15 \end{bmatrix} = (-1)(3)\det\begin{bmatrix} 1 & 2 & 4 \\ 2 & 6 & 9 \\ 1 & 2 & 5 \end{bmatrix} = (-1)(3)\det\begin{bmatrix} 1 & 2 & 4 \\ 0 & 2 & 1 \\ 0 & 0 & 1 \end{bmatrix}$$

rows 1 & 2 interchanged; row 3 has factor 3; multiples of row 1 added to rows 2 & 3

$$= (-1)(3)(1)(2)(1) = -6.$$

product of entries on main diagonal

The previous example outlines an efficient technique using elementary row transformations to evaluate the determinant of a square matrix. The procedure consists of using elementary row transformations to transform the given matrix into a triangular matrix (in the above example into an upper triangular matrix), taking into account the effect of each transformation, then finally evaluating the determinant of the resulting triangular matrix by multiplying the entries along the main diagonal.

1. Use elementary row operations to evaluate the determinants of the following matrices:

(a) $\begin{bmatrix} 2 & 3 & 4 \\ 1 & 4 & 3 \\ 3 & 6 & 9 \end{bmatrix}$ (b) $\begin{bmatrix} 3 & 1 & 5 \\ 2 & 6 & 8 \\ 1 & 3 & 2 \end{bmatrix}$ (c) $\begin{bmatrix} 4 & 8 & 4 \\ 2 & 6 & 8 \\ 3 & 9 & 6 \end{bmatrix}$ (d) $\begin{bmatrix} 1 & 2 & 3 \\ 6 & 3 & 9 \\ 4 & 2 & 8 \end{bmatrix}$

(e) $\begin{bmatrix} 2 & 3 & 5 & 4 \\ 1 & 3 & 0 & 2 \\ 2 & 4 & 6 & 4 \\ 3 & 6 & 3 & 0 \end{bmatrix}$ (f) $\begin{bmatrix} 1 & 2 & 3 & 4 \\ 1 & 4 & 4 & 7 \\ 2 & 4 & 1 & 5 \\ 3 & 6 & 9 & 2 \end{bmatrix}$ (g) $\begin{bmatrix} 3 & 1 & 2 & 4 \\ 6 & 5 & 3 & 7 \\ 9 & 3 & 2 & 1 \\ 6 & 2 & 4 & 5 \end{bmatrix}$

Properties of Determinants

Let A be a square matrix. Let $\hat{A}$ be the matrix resulting from performing one or more elementary row operation on A. Since the effect of performing an elementary row operation on the value of the determinant is either to reverse the sign or multiply the value of the determinant by a nonzero number, and since the elementary row operations are invertible operations; therefore $\det(\hat{A}) \neq 0$ if and only if $\det(A) \neq 0$ and similarly $\det(\hat{A}) = 0$ if and only if det(A) = 0.

Let a sequence of elementary row operations be performed on the $n \times n$ matrix A so as to reduce A to its reduced row-echelon form R. Now A is invertible if and only if R = I. But det (R) = det (I) = 1 ≠ 0 if and only if det (A) ≠ 0. We therefore conclude that A is invertible if and only if det(A) ≠ 0 and state this result in the form of a theorem.

Theorem: The square matrix A is invertible if and only if det (A) ≠ 0.

Example: $\det\begin{bmatrix}1 & 2\\4 & 3\end{bmatrix}=6-8=-2\neq 0$ so $\begin{bmatrix}1 & 2\\4 & 3\end{bmatrix}$ is an invertible matrix.

$\det\begin{bmatrix}1 & 2\\2 & 4\end{bmatrix}=0$ so $\begin{bmatrix}1 & 2\\2 & 4\end{bmatrix}$ is not invertible.

A direct consequence of the above theorem is the following result.

Theorem: Let A be a square matrix. Then the linear system Ax = b has a unique solution for every b if and only if det (A) ≠ 0.

Proof: Suppose det (A) ≠ 0, then A is invertible. Then $A^{-1}b$ is a solution to Ax = b since $A(A^{-1}b) = b$. To show that this is the only solution to Ax = b, suppose that xo is also a solution to Ax = b. Then $Ax_0 = b$ so $A^{-1}(A x_0) = A^{-1} b$ and hence $x_0 = A^{-1} b$. This shows that $A^{-1} b$ is the unique solution to Ax = b.

On the other hand, if Ax = b has a unique solution, then when solving this system by matrix methods the coefficient matrix is reduced to the identity matrix I and so A is invertible and hence det(A) ≠ 0.

Theorem: Let A be an n × n matrix and let E be an n × n elementary matrix.

Then, det(EA) = det(E) det(A).

Proof: The proof consists in showing that the result is true for each one of the three types of elementary matrices.

Let E be the elementary matrix obtained from I by interchanging two rows of I. Then EA is the matrix resulting from interchanging the corresponding two rows of A.

Then det(EA) = – det(A) = (–1) det(A) = det(E) det(A) since det(E) = –1.

Let E be the elementary matrix obtained from I by multiplying the entries of some row of I by a nonzero scalar k. Then EA is the matrix resulting from multiplying the entries of a row of A by k. Then det(EA) = k det(A) = det(E) det(A) since det(E) = k.

Let E be the elementary matrix obtained from I by adding a multiple of one row of I to another row of I. Then EA is the result of adding a multiple of a row of A to another row of A. Then det(EA) = det(A) = (1) det(A) = det(E) det(A) since det(E) = 1.

The result in the above theorem can be generalized to any two n × n matrices. The proof is omitted but stated in the following theorem.

Theorem: If A and B are square matrices of the same size, then det(AB) = det(A) det(B).

Example: Let $A=\begin{bmatrix}4 & 3\\1 & 2\end{bmatrix}$, and let $B=\begin{bmatrix}6 & 7\\5 & 8\end{bmatrix}$, then $AB=\begin{bmatrix}39 & 52\\16 & 23\end{bmatrix}$.

det(A) = 8 – 3 = 5 det(B) = 48 – 35 = 13 and det (AB) = 897 – 832 = 65

det(A) det(B) = (5)(13) = 65 = det(AB).

Theorem: If A is an invertible matrix, then $\det(A^{-1}) = 1/\det(A)$.

Proof: $A^{-1}A=I\Rightarrow\det\left(A^{-1}A\right)=\det(I)\Rightarrow\det\left(A^{-1}\right)\det(A)=1\Rightarrow\det\left(A^{-1}\right)=1/\det(A)$.

Theorem: If A is a square matrix, then $\det(A^T) = \det(A)$.

Proof: A cofactor expansion along the first row of A^T gives the same terms as a cofactor expansion along the first column of A.

Example: Let $A=\begin{bmatrix}2 & 1\\ 4 & 3\end{bmatrix}$, then $A^T=\begin{bmatrix}2 & 4\\ 1 & 3\end{bmatrix}$. det(A) = (2)(3) – (1)(4) = 6 – 4 = 2. det (A^T) = (2)(3) – (4)(1) = 6 – 4 = 2. So det (A) = det(A^T) = 2.

Theorem: Let A and B be square matrices with AB = I. Then BA = I.

Proof: We first show that there exists a matrix C such that CA = I and then show that in fact C = B.

Since AB = I and since $\det(I)\neq 0$, therefore $\det(A)\neq 0$. But $\det(A^T)=\det(A)$, so $\det(A^T)\neq 0$ and hence A^T is invertible. Let D denote the inverse of A^T; so $D=(A^T)^{-1}$. Then

$$A^TD=I\Rightarrow (A^TD)^T=I^T\Rightarrow D^T(A^T)^T=I^T\Rightarrow D^TA=I \text{ so } D^T=C$$

We now show C = B as follows. C = CI = C(AB) = (CA)B = IB = B.

When computing the inverse of a matrix A one should verify the correctness of the computation by demonstrating that both the products AA^{-1} and $A^{-1}A$ equal I. The preceding theorem proves that in fact that it is sufficient to show that only one of these two products needs to be shown equal to I.

The Adjoint Matrix

Recall the kth row cofactor expansion of an n × n matrix A for n ≥ 2 was defined to be $\det(A)=a_{k1}(-1)^{k+1}\det(A_{k1})+a_{k2}(-1)^{k+2}\det(A_{k2})+\cdots\cdots\cdots+a_{kn}(-1)^{k+n}\det(A_{kn})$ where the quantity $(-1)^{k+j}\det(A_{kj})$ is called the kj[th] cofactor of A. To simplify our notation we will denote this quantity by the symbol C_{kj}. Thus the kth row cofactor expansion for det(A) can be written more simply as $\det(A)=a_{k1}C_{k1}+a_{k2}C_{k2}+\cdots\cdots\cdots+a_{kn}C_{kn}$. Suppose that in this expression we replace the kth row entries $a_{k1},a_{k2}\cdots\cdots,a_{kn}$ by the jth row entries $a_{j1},a_{j2}\cdots\cdots,a_{jn}$ to get $a_{j1}C_{k1}+a_{j2}C_{k2}+\cdots\cdots+a_{jn}C_{kn}$. Such an expression would arise if the entries of the kth row of A were replaced by the entries of the jth row of A and a cofactor expansion along this new k[th] row were done. But the value of this determinant would be 0 since the matrix has two identical rows (rows k and j are same). We have thus established the following theorem.

Theorem: If A is a square matrix, then $\sum_{k=1}^{n} a_{ik}C_{jk}=\begin{cases}\det(A) & \text{if } i=j\\ 0 & \text{if } i\neq j\end{cases}$.

In a similar fashion, we can deduce the following result for a column cofactor expansion.

If a cofactor expansion along a column is used then $\sum_{k=1}^{n} a_{ki}C_{kj}=\begin{cases}\det(A) & \text{if } i=j\\ 0 & \text{if } i\neq j\end{cases}$.

Cofactor Matrix

Let A be a square matrix. The cofactor matrix of A, denoted cof(A) is the matrix obtained from A by replacing every entry of A by its cofactor.

Example:

$$\text{If } A=\begin{bmatrix}1&2&6\\5&8&7\\0&3&4\end{bmatrix},\text{ then}\operatorname{cof}(A)=\begin{bmatrix}\begin{vmatrix}8&7\\3&4\end{vmatrix}&-\begin{vmatrix}5&7\\0&4\end{vmatrix}&\begin{vmatrix}5&8\\0&3\end{vmatrix}\\-\begin{vmatrix}2&6\\3&4\end{vmatrix}&\begin{vmatrix}1&6\\0&4\end{vmatrix}&-\begin{vmatrix}1&2\\0&3\end{vmatrix}\\\begin{vmatrix}2&6\\8&7\end{vmatrix}&-\begin{vmatrix}1&6\\5&7\end{vmatrix}&\begin{vmatrix}1&2\\5&8\end{vmatrix}\end{bmatrix}=\begin{bmatrix}11&-20&15\\10&4&-3\\-34&23&-2\end{bmatrix}.$$

Adjoint Matrix

If A is a square matrix, the adjoint of A, denoted adj(A) is the transpose of the cofactor matrix; that is adj(A) = $[\operatorname{cof}(A)]^T$.

Example: Continuing with the previous example,

$$A=\begin{bmatrix}1&2&6\\5&8&7\\0&3&4\end{bmatrix},\ \operatorname{cof}(A)\begin{bmatrix}11&-20&15\\10&4&-3\\-34&23&-2\end{bmatrix}\text{ and so adj(A) }[\operatorname{cof}(A)]^T=\begin{bmatrix}11&10&-34\\-20&4&23\\15&-3&-2\end{bmatrix}$$

Consider now the product $A\cdot\operatorname{adj}(A)=\begin{bmatrix}1&2&6\\5&8&7\\0&3&4\end{bmatrix}\begin{bmatrix}11&10&-34\\-20&4&23\\15&-3&-2\end{bmatrix}=\begin{bmatrix}61&0&0\\0&61&0\\0&0&61\end{bmatrix}$

But

$$\det(A)=(1)\det\begin{bmatrix}8&7\\3&4\end{bmatrix}-(2)\det\begin{bmatrix}5&7\\0&4\end{bmatrix}+(6)\det\begin{bmatrix}5&8\\0&3\end{bmatrix}=(1)(11)-(2)(20)+(6)(15)=61.$$

We see that the product adj(A) · A is a diagonal matrix with the diagonal entries = det(A).

This example suggests the following theorem:

Theorem: If A is a square matrix, then I A · adj(A) = det(A)·

Proof: A · adj(A) =

$$\begin{bmatrix}a_{11}&a_{12}&\bullet&\bullet&a_{1n}\\a_{21}&a_{22}&\bullet&\bullet&a_{2n}\\\bullet&\bullet&\bullet&\bullet&\bullet\\\bullet&\bullet&\bullet&\bullet&\bullet\\a_{n1}&a_{n2}&\bullet&\bullet&a_{nn}\end{bmatrix}\begin{bmatrix}C_{11}&C_{21}&\bullet&\bullet&C_{n1}\\C_{12}&C_{22}&\bullet&\bullet&C_{n2}\\\bullet&\bullet&\bullet&\bullet&\bullet\\\bullet&\bullet&\bullet&\bullet&\bullet\\C_{1n}&C_{2n}&\bullet&\bullet&C_{nn}\end{bmatrix}=\begin{bmatrix}\sum_{k=1}^{n}a_{1k}C_{1k}&\sum_{k=1}^{n}a_{1k}C_{2k}&\bullet&\bullet&\sum_{k=1}^{n}a_{1k}C_{nk}\\\sum_{k=1}^{n}a_{2k}C_{1k}&\sum_{k=1}^{n}a_{2k}C_{2k}&\bullet&\bullet&\sum_{k=1}^{n}a_{2k}C_{nk}\\&\bullet&\bullet&\bullet&\bullet\\\bullet&\bullet&\bullet&\bullet&\bullet\\\sum_{k=1}^{n}a_{nk}C_{1k}&\sum_{k=1}^{n}a_{nk}C_{2k}&\bullet&\bullet&\sum_{k=1}^{n}a_{nk}C_{nk}\end{bmatrix}=$$

$$\begin{bmatrix} \det(A) & 0 & 0 & 0 & 0 \\ 0 & \det(A) & 0 & 0 & 0 \\ 0 & 0 & \det(A) & 0 & 0 \\ 0 & 0 & 0 & \det(A) & 0 \\ 0 & 0 & 0 & 0 & \det(A) \end{bmatrix} = \det(A)\begin{bmatrix} 1 & 0 & 0 & 0 & 0 \\ 0 & 1 & 0 & 0 & 0 \\ 0 & 0 & 1 & 0 & 0 \\ 0 & 0 & 0 & 1 & 0 \\ 0 & 0 & 0 & 0 & 1 \end{bmatrix} = \det(A)\cdot I$$

Theorem: If det(A) ≠ 0, then A is invertible and $A^{-1} = \frac{1}{\det(A)}\cdot \operatorname{adj}(A)$.

Proof: From the previous theorem we have A · adj(A) = det(A) · I. Since det(A) ≠ 0 we can divide by det(A) to get $A\cdot\frac{1}{\det(A)}\cdot \operatorname{adj}(A) = I$ so $A^{-1} = \frac{1}{\det(A)}\cdot \operatorname{adj}(A)$.

Example: Use the preceding theorem to find A^{-1} for the matrix $A = \begin{bmatrix} 1 & 2 & 6 \\ 5 & 8 & 7 \\ 0 & 3 & 4 \end{bmatrix}$.

Solution: This is the same matrix used in the previous example where we found,

$$\operatorname{adj}(A) = \begin{bmatrix} 11 & 10 & -34 \\ -20 & 4 & 23 \\ 15 & -3 & -2 \end{bmatrix}$$

and,

$\det(A) = 61$. Using $A^{-1} = \frac{1}{\det(A)}\operatorname{adj}(A)$ we get,

$$A^{-1} = \frac{1}{61}\begin{bmatrix} 11 & 10 & -34 \\ -20 & 4 & 23 \\ 15 & -3 & -2 \end{bmatrix} = \begin{bmatrix} 11/61 & 10/61 & -34/61 \\ -20/61 & 4/61 & 23/61 \\ 15/61 & -3/61 & -2/61 \end{bmatrix}.$$

Cramer's Rule

Cramer's rule provides a formula for solving a system of n linear equations in n variables when the system has a unique solution.

Theorem: (Cramer's Rule) Let Ax = b be a system of n linear equations in n variables with det(A) ≠ 0. Let A_k be the matrix obtained from A by replacing the kth column of A by the column vector b. Then the system has the unique solution,

$$x_k = \frac{\det(A_k)}{\det(A)}, k = 1, 2, \cdots\cdots, n.$$

Proof: Since det(A) ≠ 0, A is invertible and the system Ax = b has the unique solution:

$$x = A^{-1}b.$$

Therefore,

$$x = \frac{1}{\det(A)}\operatorname{adj}(A)\cdot b = \frac{1}{\det(A)}\begin{bmatrix} C_{11} & C_{21} & \bullet & \bullet & C_{n1} \\ C_{12} & C_{22} & \bullet & \bullet & C_{n2} \\ \bullet & \bullet & \bullet & \bullet & \bullet \\ \bullet & \bullet & \bullet & \bullet & \bullet \\ C_{1n} & C_{2n} & \bullet & \bullet & C_{nn} \end{bmatrix}\begin{bmatrix} b_1 \\ b_2 \\ \bullet \\ \bullet \\ b_n \end{bmatrix}$$

$$= \frac{1}{\det(A)}\begin{bmatrix} b_1C_{11} + b_2C_{21} + \cdots\cdots\cdots + b_nC_{n1} \\ b_1C_{1n} + b_2C_{2n} + \cdots\cdots\cdots + b_nC_{nn} \\ \bullet \\ \bullet \\ b_1C_{1n} + b_2C_{2n} + \cdots\cdots\cdots + b_nC_{nn} \end{bmatrix}$$

Thus, $x_k = \dfrac{b_1C_{1k} + b_2C_{2k} + \cdots\cdots\cdots + b_nC_{nk}}{\det(A)}$ for $k = 1, 2, \cdots\cdots\cdots, n$.

Now the numerator of x_k consists of the kth column cofactors of A multiplied by the corresponding entries of b. We get the same result if we use a kth column cofactor expansion of A_k so $x_k = \dfrac{\det(A_k)}{\det(A)}$ for $k = 1, 2 \cdots\cdots\cdots, n$.

Example: Use Cramer's rule to solve the following system of linear equations.

$$x + 2y + z = 9$$
$$2x + y + z = 7$$
$$x + y + 3z = 10$$

Solution:

$$x = \frac{\det\begin{bmatrix} 9 & 2 & 1 \\ 7 & 1 & 1 \\ 10 & 1 & 3 \end{bmatrix}}{\det\begin{bmatrix} 1 & 2 & 1 \\ 2 & 1 & 1 \\ 1 & 1 & 3 \end{bmatrix}} = \frac{-7}{-7} = 1 \quad y = \frac{\det\begin{bmatrix} 1 & 9 & 1 \\ 2 & 7 & 1 \\ 1 & 10 & 3 \end{bmatrix}}{\det\begin{bmatrix} 1 & 2 & 1 \\ 2 & 1 & 1 \\ 1 & 1 & 3 \end{bmatrix}} = \frac{-21}{-7} = 3 \quad z = \frac{\det\begin{bmatrix} 1 & 2 & 9 \\ 2 & 1 & 7 \\ 1 & 1 & 10 \end{bmatrix}}{\det\begin{bmatrix} 1 & 2 & 1 \\ 2 & 1 & 1 \\ 1 & 1 & 3 \end{bmatrix}} = \frac{-14}{-7} = 2$$

1. Use Cramer's rule to solve the following systems of linear equations.

(a) $\begin{matrix} 2x + 3y = 8 \\ x - 2y = -3 \end{matrix}$ (b) $\begin{matrix} 4x + y = 13 \\ x + 3y = 6 \end{matrix}$ (c) $\begin{matrix} x + 2y = 1 \\ 2x + y = 5 \end{matrix}$

$(a)\begin{matrix}2x+3y=8\\x-2y=-3\end{matrix}$ $(b)\begin{matrix}4x+y=13\\x+3y=6\end{matrix}$ $(c)\begin{matrix}x+2y=1\\2x+y=5\end{matrix}$

Applications using the Determinant

Two applications using the determinant will be considered. The first is a convenient method for evaluating a cross product. The second application is finding the volume of a parallelepiped.

Cross Product

Recall if $u=(u_1,u_2,u_3)$ and $v=(v_1,v_2,v_3)$, then,

$$\begin{aligned}u\times v&=(u_2v_3-u_3v_2,u_3v_1-u_1v_3,u_1v_2-u_2v_1)\\&=(u_2v_3-u_3v_2)i+(u_3v_1-u_1v_3)j+(u_1v_2-u_2v_1)k\end{aligned}$$

where i = (1, 0, 0), j = (0, 1, 0) and k = (0, 0, 1).

A convenient method for arriving at the cross product u × v is to use the matrix form $\begin{bmatrix}i & j & k\\u_1 & u_2 & u_3\\v_1 & v_2 & v_3\end{bmatrix}$ and do a first row cofactor expansion as you would do to evaluate a determinant.

Example: Let u = (3, 2, 1) and let v = (4, 5, 6). Find u × v.

Solution: $u\times v=\begin{vmatrix}i & j & k\\3 & 2 & 1\\4 & 5 & 6\end{vmatrix}=\begin{vmatrix}2 & 1\\5 & 6\end{vmatrix}i=\begin{vmatrix}3 & 1\\4 & 6\end{vmatrix}j=\begin{bmatrix}3 & 2\\4 & 5\end{bmatrix}k=7i-14j+7k=(7,-14,7)$

Volume of a Parallelepiped

Recall that if a parallelepiped has the vectors u, v and w as edges, then the volume of the parallelepiped is $|w\cdot(u\times v)|$. A straight forward calculation shows that

$$w\cdot(u\times v)=\det\begin{bmatrix}w_1 & w_2 & w_3\\u_1 & u_2 & u_3\\v_1 & v_2 & v_3\end{bmatrix}=\det\begin{bmatrix}u_1 & u_2 & u_3\\v_1 & v_2 & v_3\\w_1 & w_2 & w_3\end{bmatrix}$$

Hence the volume of the parallelepiped can be found by taking the absolute value of either of the above determinants.

Example: Find the volume of the parallelepiped having the vectors (1, 0, 1), (2, 3, 4) and (3, 5, 5) as edges.

Solution:

$$\text{Volume}=\left|\det\begin{bmatrix}1 & 0 & 1\\2 & 3 & 4\\3 & 5 & 5\end{bmatrix}\right|=|-4|=4.$$

1. Calculate the following cross products.

- (2,3,1)× (3,1,4).
- (1,0,2)× (3,5,4).
- (5,3,4)× (2,0,1).
- (3,0,1) (2,5,4).

2. Find the volume of the parallelepiped having the following vectors as edges.

- (2, 3, 1), (3, 4, 6), (5, 7, 9).
- (1, 0, 2), (3, 5, 7), (5, 6, 8).
- (2, 0, 1), (5, 4, 6), (7, 8, 8).
- (4, 6, 5), (2, 1, 3), (7, 8, 7).

References

- Adjoint-and-inverse-of-a-matrix: byjus.com, Retrieved 12, April 2020
- Rank-of-a-matrix-and-special-matrices: byjus.com, Retrieved 30, July 2020
- Determinants: cc.umanitoba.ca, Retrieved 15, March 2020

5

Experimental Design

The design of any task which seeks to describe and explain the variation of information within conditions which are hypothesized to reflect the variation is known as experimental design. This chapter closely examines the key concepts of experimental design, such as completely randomized design and randomized block and Latin square design, to provide an extensive understanding of the subject.

Basics of Experimental Design

Design of experiment means how to design an experiment in the sense that how the observations or measurements should be obtained to answer a query in a valid, efficient and economical way. The designing of the experiment and the analysis of obtained data are inseparable. If the experiment is designed properly keeping in mind the question, then the data generated is valid and proper analysis of data provides the valid statistical inferences. If the experiment is not well designed, the validity of the statistical inferences is questionable and may be invalid.

It is important to understand first the basic terminologies used in the experimental design.

- Experimental Unit: For conducting an experiment, the experimental material is divided into smaller parts and each part is referred to as an experimental unit. The experimental unit is randomly assigned to treatment is the experimental unit. The phrase "randomly assigned" is very important in this definition.
- Experiment: A way of getting an answer to a question which the experimenter wants to know.
- Treatment: Different objects or procedures which are to be compared in an experiment are called treatments.
- Sampling Unit: The object that is measured in an experiment is called the sampling unit. This may be different from the experimental unit.
- Factor: A factor is a variable defining a categorization. A factor can be fixed or random in nature. A factor is termed as a fixed factor if all the levels of interest are included in the experiment.

  A factor is termed as a random factor if all the levels of interest are not included in the experiment and those that are can be considered to be randomly chosen from all the levels of interest.
- Replication: It is the repetition of the experimental situation by replicating the experimental unit.

- Experimental Error: The unexplained random part of the variation in any experiment is termed as experimental error. An estimate of experimental error can be obtained by replication.
- Treatment Design: A treatment design is the manner in which the levels of treatments are arranged in an experiment.

Suppose some varieties of fish food is to be investigated on some species of fishes. The food is placed in the water tanks containing the fishes. The response is the increase in the weight of fish. The experimental unit is the tank, as the treatment is applied to the tank, not to the fish. Note that if the experimenter had taken the fish in hand and placed the food in the mouth of fish, then the fish would have been the experimental unit as long as each of the fish got an independent scoop of food.

Design of Experiment

One of the main objectives of designing an experiment is how to verify the hypothesis in an efficient and economical way. In the contest of the null hypothesis of equality of several means of normal populations having the same variances, the analysis of variance technique can be used. Note that such techniques are based on certain statistical assumptions. If these assumptions are violated, the outcome of the test of a hypothesis then may also be faulty and the analysis of data may be meaningless. So the main question is how to obtain the data such that the assumptions are met and the data is readily available for the application of tools like analysis of variance. The designing of such a mechanism to obtain such data is achieved by the design of the experiment. After obtaining the sufficient experimental unit, the treatments are allocated to the experimental units in a random fashion. Design of experiment provides a method by which the treatments are placed at random on the experimental units in such a way that the responses are estimated with the utmost precision possible.

Principles of Experimental Design

There are three basic principles of design which were developed by Sir Ronald A. Fisher.

- Randomization.
- Replication.
- Local control.

Randomization

The principle of randomization involves the allocation of treatment to experimental units at random to avoid any bias in the experiment resulting from the influence of some extraneous unknown factor that may affect the experiment. In the development of analysis of variance, we assume that the errors are random and independent. In turn, the observations also become random. The principle of randomization ensures this.

The random assignment of experimental units to treatments results in the following outcomes:

- It eliminates systematic bias.

- It is needed to obtain a representative sample from the population.
- It helps in distributing the unknown variation due to confounded variables throughout the experiment and breaks the confounding influence.

Randomization forms a basis of a valid experiment but replication is also needed for the validity of the experiment.

If the randomization process is such that every experimental unit has an equal chance of receiving each treatment, it is called complete randomization.

Replication

In the replication principle, any treatment is repeated a number of times to obtain a valid and more reliable estimate than which is possible with one observation only. Replication provides an efficient way of increasing the precision of an experiment. The precision increases with the increase in the number of observations. Replication provides more observations when the same treatment is used, so it increases precision. For example, if the variance of x is σ^2 than variance of the sample mean $\bar{x}$ based on n observation is $\frac{\sigma^2}{n}$. So as n increases, $\text{var}(\bar{x})$ decreases.

Local Control or Error Control

The replication is used with local control to reduce the experimental error. For example, if the experimental units are divided into different groups such that they are homogeneous within the blocks, then the variation among the blocks is eliminated and ideally, the error component will contain the variation due to the treatments only. This will, in turn, increase the efficiency.

Complete and Incomplete Block Designs

In most of the experiments, the available experimental units are grouped into blocks having more or less identical characteristics to remove the blocking effect from the experimental error. Such design is termed as block designs.

The number of experimental units in a block is called the block size.

If size of block = number of treatments and each treatment in each block is randomly allocated, then it is a full replication and the design is called a complete block design.

In case, the number of treatments is so large that a full replication in each block makes it too heterogeneous with respect to the characteristic under study, then smaller but homogeneous blocks can be used. In such a case, the blocks do not contain a full replicate of the treatments. Experimental designs with blocks containing an incomplete replication of the treatments are called incomplete block designs.

Construction of Experimental Design

Before the construction of experimental design it is essential for the experimenter to have a clear idea about the objective in advance what is our objective, how to collect the data and how to do analyses of this data.

(a) First to recognize the problem: It is essential to first understand the problem and make a valid statement about the objective. It may seem a very simple task at first approach but it is not. As it require a lot brain storming sessions to develop a clear objective of the experiment. Sometime it is essential to take input from all the concerned specialized persons about the system like engineers, marketing, management, customer and operators. Hence for designing the experiment the collective efforts of the group is required.

It is helpful if we prepare a list of questions and problems that are raised by the experimenter. The clear statement about the objective will help for better understanding of the system and the solution to the problem.

It is also essential to keep the overall objective in mind while performing an experiment. There are many reasons for conducting the experiments and each experiment will generate its own series of questions. Reasons for conducting an experiment are given as:

- The main objective of conducting an experiment is to determine or find out the major factors that have the most influence on the response variable. The process of finding the factors that are important is called factor screening.
- After determining the factors that the next step is to determine the optimal stage or level that will give you the optimal values of the response. For example, while experimentation of the yield of the crop it is determined that quality of seeds, fertility of land and fertilizers used are the main factors. Then the next stage is to find the optimum combinations of all these to get the best results in terms of increased profit or minimized cost.
- Now after determining the levels the factors, the experimenter is trying to verify that the system operates in the same manner as previously or there are some changes in the structure of the system with time.
- It may also happen that the experiment are trying to apply new resources to see their impact on the system. For example, in manufacturing, experiments are conducted with new resources that increase the quality as well as the reliability of the response. Experimentation of this type will increase the demand of the product and overall profit.
- The experimentation will also help to test the robustness of the response variable. Under what conditions the value of the response variable will deviate from the target value.

(b) As we have already determined the factors those are the most influential for the response variable. Now the next step is to select the response variable, the experimenter should be certain about the response variable that will give the useful information about the system under study.

(c) After identifying the factors, the next important task is to categorize these factors into design factors, factors that are constant and factors that can vary. The design factors are those factors that are selected for study in the experiment. Constant factors have no direct impact for the current experiment but they can affect the system in future. Varying factors are those factors that have property of inhomogeneous among them for example, the experimental units or materials are not completely homogenous. By using the principle of randomization, one can reduce the effect of this type of variability on the response variable.

There are many other factors that have major influence on the response variable. These factors are called nuisance factors. Nuisance factors are classified as controllable (assignable) and (non-assignable) uncontrollable. Controllable nuisance factors are those whose factor's levels are set by the experimenter. On the other hand, if the factors are uncontrollable but still measurable then the procedure that is used for analysis is called analysis of covariance.

(d) If we are able to categorize our factors then the next step is to choose the design according to the objective. It involves determination of the sample size as well as number of replicates, determination of whether to involve blocking or not in the design. There are many statistical software available that will help to fit the design the model on the data. Based on the information in terms of number of factors and their levels software will recommend a particular model. It is also possible that one can compare between models based on the coefficient of determination as well as to give some diagnostic information about the performance of the model. This will help the experimenter to determine the suitable model for the objective. One can start from the very simple model and add the factors/ variables in the model based on the requirements of the objective and attain the optimal model.

(e) The most important phase of designing experiment is the running stage of the experiment. It is important that process should perform according to the plan. If the process is not working about to the plan then it will bring serious consequences like wastage of material, time, human resources etc. So it is suggested that one should perform some pilot study or runs before implementing the experiment on the large scale. These runs will help to know about the consistency about the experimental design. If it is not working properly, then there is a need to cross check all the previous stage and again do pilot study.

(f) After running the experiment, the next step is to collect the data to apply statistical methods on them. There are many software available that will help to assist in analysis. Descriptive statistics play an important role in data analysis at initial stage but most the studies require methods of inferential statistics like estimation of variable coefficient, hypothesis testing and confidence interval. Residual analysis and model adequacy checking are among the major analysis techniques.

(g) After analysis the data using the appropriate statistical methods, the next stage is to provide an appropriate conclusions about the results. One can also test the validity of the conclusion by performing confirmation testing.

Hence, as we have observed the importance of experimentation and steps how to implement it. One has to work very efficiently stating from the initial stage of determining the factors till the conclusion to meet out the objective.

Use of Statistical Techniques in Experimentation

The basic purpose of the use of statistical techniques in experimentation is to increase the efficiency of these experiments and to provide a valid conclusions. The following points should be kept in mind by the experimenter while applying the statistical techniques:

- The most important thing that the experimenter should have a strong knowledge in their field. As this knowledge will help in determining the factors that are most influential for the response variable and also in the designing of the model.
- Second important thing that the experimenter should keep in mind that one should start with simple statistical techniques and keep the design as simple as simple. As for a simple and reasonable design the analysis will be very simple and it is easy to provide its interpretations.
- There are situations when there is a strong need to interpret the statistical results according to practical situation. For example, if there is a statistical significant difference in the yield of crop by using two quality of seeds then one should consider the significant difference in terms of other factors like cost of production new seed than the old one as well as the total benefit that one will receive while using the new one. If the total profit that one will earn is negligible then however there is a significant difference in terms of statistics but in practical case there is no difference.
- It is possible that to draw a valid conclusion about the system one has to repeat the experimentation many times. As each experiment collects some information so one can use this information to get the better results in further experiments.

Completely Randomized Design

Description of the Design

- Simplest design to use.
- Design can be used when experimental units are essentially homogeneous.
- Because of the homogeneity requirement, it may be difficult to use this design for field experiments.
- The CRD is best suited for experiments with a small number of treatments.

Randomization Procedure

- Treatments are assigned to experimental units completely at random.
- Every experimental unit has the same probability of receiving any treatment.
- Randomization is performed using a random number table, computer, program, etc.

Example of Randomization

Given you have 4 treatments (A, B, C, and D) and 5 replicates, how many experimental units would you have?

1	2	3	4	5	6	7	8	9	10
D	D	B	C	D	C	A	A	B	D
11	12	13	14	15	16	17	18	19	20
C	B	A	B	C	B	C	D	A	A

- Note that there is no "blocking" of experimental units into replicates.
- Every experimental unit has the same probability of receiving any treatment.

Advantages of a CRD

- Very flexible design (i.e. number of treatments and replicates is only limited by the available number of experimental units).
- Statistical analysis is simple compared to other designs.
- Loss of information due to missing data is small compared to other designs due to the larger number of degrees of freedom for the error source of variation.

Disadvantages of a CRD

- If experimental units are not homogeneous and you fail to minimize this variation using blocking, there may be a loss of precision.
- Usually the least efficient design unless experimental units are homogeneous.
- Not suited for a large number of treatments.

Fixed vs. Random Effects

- The choice of labeling a factor as a fixed or random effect will affect how you will make the F-test.
- This will become more important later in the course when we discuss interactions.

Fixed Effect

- All treatments of interest are included in your experiment.
- You cannot make inferences to a larger experiment.

Example 1: An experiment is conducted at Fargo and Grand Forks, ND. If location is considered a fixed effect, you cannot make inferences toward a larger area (e.g. the central Red River Valley).

Example 2: An experiment is conducted using four rates (e.g. ½ X, X, 1.5 X, 2 X) of a herbicide to determine its efficacy to control weeds. If rate is considered a fixed effect, you cannot make inferences about what may have occurred at any rates not used in the experiment (e.g. ¼ x, 1.25 X, etc.).

Random Effect

- Treatments are a sample of the population to which you can make inferences.
- You can make inferences toward a larger population using the information from the analyses.

Example 1: An experiment is conducted at Fargo and Grand Forks, ND. If location is considered a random effect, you can make inferences toward a larger area (e.g. you could use the results to state what might be expected to occur in the central Red River Valley).

Example 2: An experiment is conducted using four rates (e.g. ½ X, X, 1.5 X, 2 X) of an herbicide to determine its efficacy to control weeds. If rate is considered a random effect, you can make inferences about what may have occurred at rates not used in the experiment, e.g. ¼ x, 1.25 X, etc.

Analysis of the Fixed Effects Model

Notation

- Statistical notation can be confusing, but use of the Y-dot notation can help simplify things.
- The dot in the Y-dot notation implies summation across over the subscript it replaces.
- For example,
    - $y_{i} = \sum_{j=1}^{n} y_{ij}$ = Treatment total, where n = number of observation in a treatment.
    - $\bar{y}_{i} = y_{i}/n$ = Treatment mean.
    - $y_{..} = \sum_{i=1}^{a} \sum_{j=1}^{n} y_{ij}$ = Experiment total, where a = number of treatments.
    - $\bar{y}_{..} = y_{..}/N$ = n, Experiment, where N = total number of observation in the experiment.

Linear Additive Model for the CRD

$$Y_{ij} = \mu + \tau_{i} + \varepsilon_{ji}$$

where: Y_{ij} is the jth observation of the ith treatment,

μ is the population mean,

τ_{i} is the treatment effect of the ith treatment, and

ε_{ji} is the random error.

Using this model we can estimate τ_{i} or ε_{ji} for any observation if we are given Y_{ij} and μ.

Example:

	Treatment 1	Treatment 2	Treatment 3	
	4 5 6	9 10 11	8 11 8	
Y_i	15	30	27	$Y_{..} = 72$
$\bar{Y}_{i.}$	5	10	9	$\bar{Y}_{..} = 8$
$\bar{Y}_{i.} - \bar{Y}_{..}$	-3	2	1	

We can now write the linear model for each observation (Y_{ij}).

Write in μ for each observation.

	Treatment 1	Treatment 2	Treatment 3	
	4 = 8 5 = 8 6 = 8	9 = 8 10 = 8 11 = 8	8 = 8 11 = 8 8 = 8	
Y_i	15	30	27	$Y_{..} = 72$
$\bar{Y}_{i.}$	5	10	9	$\bar{Y}_{..} = 8$
$\bar{Y}_{i.} - \bar{Y}_{..}$	-3	2	1	

Write in the respective τ_i for each observation where $\tau_i = \bar{Y}_{i.} - \bar{Y}_{..}$

	Treatment 1	Treatment 2	Treatment 3	
	4 = 8 − 3 5 = 8 − 3 6 = 8 − 3	9 = 8 + 2 10 = 8 + 2 11 = 8 + 2	8 = 8 + 1 11 = 8 + 1 8 = 8 + 1	
Y_i	15	30	27	$Y_{..} = 72$
$\bar{Y}_{i.}$	5	10	9	$\bar{Y}_{..} = 8$
$\bar{Y}_{i.} - \bar{Y}_{..}$	-3	2	1	

Write in the ε_{ji} for each observation.

	Treatment 1	Treatment 2	Treatment 3	
	4 = 8 − 3 -1 5 = 8 − 3 + 0 6 = 8 − 3 + 1	9 = 8 + 2 - 1 10 = 8 + 2 + 0 11 = 8 + 2 + 1	8 = 8 + 1 - 1 11 = 8 + 1 + 2 8 = 8 + 1 - 1	

Y_{i}	15	30	27	$Y_{..} = 72$
$\overline{Y}_{i.}$	5	10	9	$\overline{Y}_{..} = 8$
$\overline{Y}_{i.} - \overline{Y}_{..}$	-3	2	1	

Note for each treatment $\sum \varepsilon_{ij} = 0$.

Examples: If you are given just the treatment totals ($Y_{i.}$'s), how would you fill in the Yi. values for each of the observations such that the Error SS = 0.

Solution: Remember that the Experimental Error is the failure of observations treated alike to be the same. Therefore, if all treatments have the same value in each replicate, the Experimental Error SS =0.

Example: Given the following information, fill in the values for all Y_{ij}'s such that the Experimental Error SS = 0.

	Treatment 1	Treatment 2	Treatment 3	
Y_{i}	15	30	27	$Y_{..} = 72$
$\overline{Y}_{i.}$	5	10	9	$\overline{Y}_{..} = 8$

Solution:

	Treatment 1	Treatment 2	Treatment 3	
	5 5 5	10 10 10	9 9 9	
Y_{i}	15	30	27	$Y_{..} = 72$
$\overline{Y}_{i.}$	5	10	9	$\overline{Y}_{..} = 8$

The previous two examples that $\sum \tau_i = 0$. This is true for all situations.

Given,

$$H_0 : \mu_1 = \mu_2 = \ldots = \mu_t$$

($H_A : \mu_i \neq \mu_1$) for at least one pair of treatments (i, i').

$\frac{\sum_{i=1}^{t} \mu_i}{t} = \mu$ (i.e., sum of the treatment means divide by the number of treatments equal the experiment mean.)

- This definition implies that $\sum_{i=1}^{t} \tau_i = 0 = \left(\overline{Y}_i - \overline{Y}_{..}\right)$.

- The hypothesis written above can be rewritten in terms of the treatment effects τ_i as:

  $H_o : \tau_1 = \tau_2 = \ldots = \tau_a = 0$

  $H_A : \tau_i \neq 0$

- Thus, when we are testing the null hypothesis that all treatments means are the same, we are testing at the same time the null hypothesis that all treatment effects, τ_i, are zero.

Partitioning the Total Sum of Squares

- Remember that:

  $\mu = \overline{Y}$

  $\tau_i = \left(\overline{Y}_i - \overline{Y}_{..}\right)$

  $\varepsilon_{ij} = \left(Y_{ij} - \overline{Y}_i\right)$

- Thus, $Y_{ij} = \mu + \tau_i + \varepsilon_{ij}$ can be rewritten as: $Y_{ij} = \overline{Y} + \left(\overline{Y}_i - \overline{Y}_{..}\right) + \left(Y_{ij} - \overline{Y}_i\right)$.
- The Analysis of Variance is derived from the partitioning of the corrected Total Sum of Squares.

Total sum of squares = $\sum_{i=1}^{t}\sum_{j=1}^{r}\left(Y_{ij} - \overline{Y}_{..}\right)^2$

and

$$\sum_{i=1}^{t}\sum_{j=1}^{r}\left(Y_{ij} - \overline{Y}_{..}\right)^2 = \sum_{i=1}^{a}\sum_{j=1}^{n}\left[\left(\overline{Y}_{i.} - \overline{Y}_{..}\right) + \left(Y_{ij} - \overline{Y}_i\right)\right]^2$$

$$= r\sum_{i=1}^{t}\left(\overline{Y}_{i.} - \overline{Y}_{..}\right)^2 + \sum_{i=1}^{t}\sum_{j=1}^{r}\left(\overline{Y}_{ij} - \overline{Y}_{..}\right)^2 + 2\sum_{i=1}^{t}\sum_{j=1}^{r}\left(\overline{Y}_{i.} - \overline{Y}\right)\left(Y_{ij} - \overline{Y}_i\right)$$

The last term of the equation equals zero because $\sum_{j=1}^{r}\left(Y_{ij} - \overline{Y}_{i.}\right) = \sum \varepsilon_{ij} = 0$.

Thus, $\sum_{i=1}^{t}\sum_{j=1}^{r}\left(Y_{ij} - \overline{Y}_{..}\right)^2 + r\sum_{i=1}^{t}\left(\overline{Y}_{i.} - \overline{Y}_{..}\right)^2 + \sum_{i=1}^{t}\sum_{j=1}^{r}\left(Y_{ij} - \overline{Y}_{..}\right)^2$, which is,

Total Sum of Squares = Treatment Sum of Squares + Error Sum of Squares

ANOVA for Any Number of Treatments with Equal Replication

	Treatment			
Replicate	A	B	C	
1	23	42	47	
2	36	26	43	
3	31	47	43	
4	33	34	39	

Y_i	123	149	172	Y...=444
$\sum Y_{ij}^2$	3,875	5,805	7,428	

Step 1. Write the hypotheses to be tested.

$$H_o : \mu_1 = \mu_2 = \mu_3$$
$$H_A : \mu_i = \mu_2 \neq \mu_3$$

Or

$$\mu_1 \neq \mu_2 = \mu_3$$

Or

$$\mu_1 \neq \mu_2 \neq \mu_3$$

- H_o : All three means are equal.
- H_A : At least one of the means is different from the other means.

Step 2. Calculate the Correction Factor.

$$CF - \frac{Y^2}{rt} - \frac{444^2}{4*3} - 16,428.0$$

Step 3. Calculate the Total SS.

$$totalSS = \sum Y_{ij}^2 - CF$$
$$= \left(23^2 + 36^2 + 31^2 + ... + 39^2\right) - CF$$
$$= 17,108 - 16,428$$
$$= 680..0$$

Step 4. Calculate the Treatment SS (TRT SS).

$$TRTSS = \sum \frac{Y_{i.}^2}{r} - CF$$
$$= \left(\frac{123^2}{4} + \frac{149^2}{4} + \frac{172^2}{4}\right) - 16,428$$
$$= 16728.5 - 16428.0$$
$$= 300.5$$

Step 5. Calculate the Error SS.

Error SS = Total SS – Treatment SS

= 680 – 300.5

= 379.5

Step 6. Complete the ANOVA table.

Sources of variation	Df	SS	MS	F
Treatment	t-1 = 2	300.5	150.25	3.563NS
Error	t(r-1) = 9	379.5	42.167	
Total	rt-1 = 11	680.0		

Step 7. Look up Table F-values.

$F_{0.05;2,9} = 4.26$

$F_{0.01;2,9} = 8.02$

Step 8. Make conclusions.

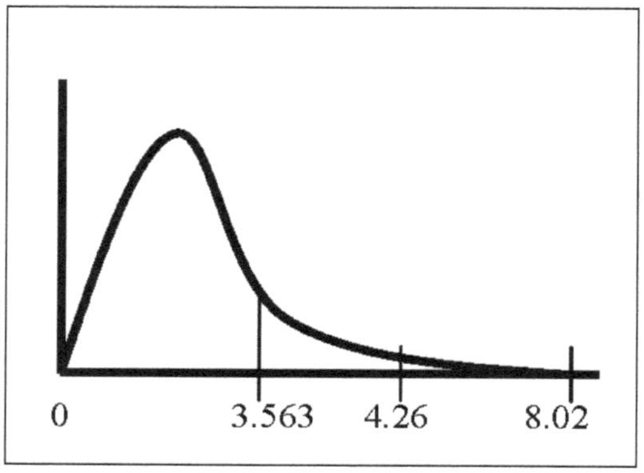

- Since F-calc (3.563) < 4.26 we fail to reject Ho: $\mu_1 = \mu_2 = \mu_3$ at the 95% level of confidence.
- Since F-calc (3.563) < 8.02 we fail to reject Ho: $\mu_1 = \mu_2 = \mu_3$ at the 99% level of confidence.

Step 9. Calculate Coefficient of Variation (CV).

$$\%CV = \frac{s}{\bar{\bar{Y}}} * 100$$

Remember that the Error MS = s^2.

$$\%CV = \frac{\sqrt{0.051}}{\left(\frac{35.5}{17}\right)} * 100$$

$$= (0.2259/2.088) * 100$$

$$= 10.82\%$$

ANOVA for Any Number of Treatments with Unequal Replication

Linear Mode $Y_{ijk} = \mu + \tau_i + \varepsilon_{ij} + \delta_{ijk}$

Where: Y_{ijk} is the kth sample of the jth observation of the ith treatment,

μ is the population mean,

τ_i is the treatment effect of the ith treatment,

ε_{ij} is the random error, and

δ_{ijk} is the sampling error.

ANOVA table:

SOV	Df	F
Treatment	t-1	Treatment MS/Experimental Error MS
Experimental error	(tr-1) - (t-1)	
Sampling Error	(trs-1) - (tr-1)	
Total	trs-1	

Facts about ANOVA with Sampling

There are two sources of variation that contribute to the variance appropriate to comparisons among treatment means:

- Sampling Error = variation among sampling units treated alike $\left(\sigma_s^2\right)$.
- Experimental Error = variation among experimental units treated alike $\left(\sigma_s^2 + r\sigma_E^2\right)$.
- The Experimental Error MS is expected to be larger than the Sampling Error MS.
- If the Experimental Error variance component is not important, the Sampling Error MS and the Experimental Error MS will be of the same order of magnitude.
- If the Experimental Error variance component is important, the Experimental Error MS will be much larger than the Sampling Error MS.

Example:

Temperature									
	8°			12°			16°		
	Pot Number			Pot Number			Pot Number		
Plant	1	2	3	1	2	3	1	2	3
1	3.5	2.5	3.0	5.0	3.5	4.5	5.0	5.5	5.5
2	4.0	4.5	3.0	5.5	3.5	4.0	4.5	6.0	4.5
3	3.0	5.5	2.5	4.0	3.0	4.0	5.0	5.0	6.5
4	4.5	5.0	3.0	3.5	4.0	5.0	4.5	5.0	5.5
$Y_{ij.}$	15.0	17.5	11.5	18.0	14.0	17.5	19.0	21.5	22.0
$Y_{i.}$		44.0			49.5			62.5	

Y...=156.0

Note i = treatment, j = replicate, and k = sample.

Step 1. Calculate correction factor:

$$\frac{Y^2_{...}}{rts} = \frac{156^2}{3(3)(4)} = 676$$

Step 2. Calculate the Total SS:

$$\text{Total SS} = \sum Y^2_{ijk} - CF$$

$$\begin{aligned} &= \left(3.5^2 + 4.0^2 + 3.0^2 + + ... + 5.5^2\right) - CF \\ &= 712.5 - 676.0 \\ &= 36.5 \end{aligned}$$

Step 3. Calculate the Treatment SS:

$$\text{Treatment SS} = \sum \frac{Y^2_{i..}}{rs} - CF$$

$$\begin{aligned} &= \left(\frac{44^2}{3(4)} + \frac{49.5^2}{(3(4)} + \frac{62.5^2}{3(4)}\right) - 676.0 \\ &= 691.04 - 676.0 \\ &= 15.042 \end{aligned}$$

Step 4. Calculate the SS Among Experimental Units Total (SSAEUT):

$$\text{SSAEUT} = \sum \frac{Y^2_{ij..}}{rs} - CF$$

$$\begin{aligned} &= \left(\frac{15^2}{4} + \frac{17.5^2}{4} + \frac{11.5^2}{4} + ... + \frac{22.0^2}{4}\right) - 676.0 \\ &= 699.25 - 676.0 \\ &= 23.25 \end{aligned}$$

Step 5. Calculate the Experimental Error SS:

Experimental Error SS = SSAEUT – SS TRT

= 23.25 – 15.042

= 8.208

Step 6. Calculate the Sampling Error SS:

Sampling Error SS = Total SS – SSAEUT

= 36.5 – 23.25

= 13.25

Step 7. Complete the ANOVA table:

SOV	Df	SS	MS	F
Treatment	t-1=2	15.042	7.521	5.498*
Experimental error	(tr-1) – (t-1) =6	8.208	1.368	
Sampling Error	(trs-1) - (tr-1) = 27	13.25		
Total	trs-1=35	36.5		

Step 8. Look up table F-values:

$F_{0.05;2,6} = 5.14$

$F_{0.01;2,6} = 10.92$

Step 9. Make conclusions:

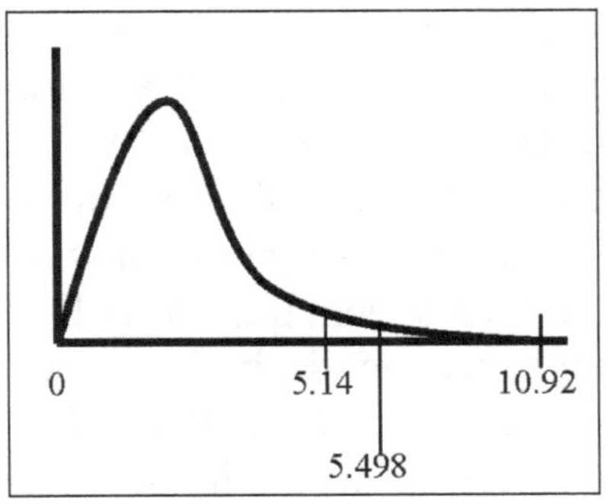

- Since F-calc (5.498) > 5.14 we reject H_0: at the 95% level of confidence.
- Since F-calc (5.498) < 10.92 we fail to reject Ho: at the 99% level of confidence.

ANOVA When the Number of Subsamples are Not Equal

$$\text{Total SS} = \sum Y_{ijk}^2 - \frac{Y_{...}^2}{\text{total \# of observation}} \qquad df = \#\text{observation} - 1$$

$$\text{Treatments SS} = \sum \frac{Y_{i..}^2}{r_j s_k} - \frac{Y_{...}^2}{\text{total \# of obs.}} \qquad df = \#\text{treatments} - 1$$

$$\text{SSAEUT} = \sum \frac{Y_{ij..}^2}{s_k} - \frac{Y_{...}^2}{\text{total \# of obs.}} \qquad df = \#\text{Experiment units} - 1$$

SS Experimental Error = SSAEUT – SS TRT   df = SSAEUT df – TRT df

SS Sampling Error = Total SS – SSAEUT   df = Total df – SSAEUT df

Assumptions Underlying ANOVA

- Experimental errors are random, independently, and normally distributed about a mean of zero and with a common variance (i.e. treatment variances are homogenous).
- The above assumption can be express as NID $(0, \sigma^2)$.

- Departure from this assumption can affect both the level of significance and the sensitivity of F- or t-tests to real departures from H_0.
- This results in the rejection of H_0 when it is true (i.e. a Type I Error) more often than α calls for.
- The power of the test also is reduced if the assumption of NID (0, σ^2) with the fixed model is usually of little consequence because ANOVA is a very robust technique.
- Violation of the basic assumptions of ANOVA can be investigated by observing plots of the residuals.
- Residuals will be discussed in more detail when Transformations are discussed later in the semester.

Adequacy of the Model

We will discuss about the validity of the assumptions before applying the ANOVA. We have to check the assumption of normality of the error term. There are methods like histogram, dot plot, line chart, Q-Q plot, Cumulative Frequency (P-P) plots and through test Kolmogorov- Smirnov (KS) test, W/S test, Jarque-Bera test, Shapiro-Wilks test and D'Agostino test one can determine the normality of the error term. Graphical representation will not be the exact method to check about the normality of the data if the values seem like normally distributed but they are not actually normal. Graphical methods are not very useful when the sample size is very small. Then the histogram of the data does not look normal but in reality it may follow normal distribution. Statistical tests give more actual probabilities about the normality of the data. In the testing of hypothesis we want to test about whether the data was drawn from the normal population or not.

The hypotheses to test about the normality of the data are:

- H_0: The sample data are taken from the normal distribution or the sample data are not significantly different from the normal population.
- H_1: The sample data are not from the normal distribution or the sample data are significantly different from the normal distribution.

There is a difference in testing of this hypothesis then others where our main motive is to estimate about the difference in mean, median and variance of the data.

In these types of hypothesis of differences, our interest in to reject the null hypothesis. If the probability value (p-value) of these events is coming out to be less than 5% then we reject the null hypothesis. Here in the testing for the normality of the data, our main interest is to see our p-value to be greater than 5% so that we can accept our null hypothesis.

So if probability value > 0.05 then it mean that the data follow normal distribution otherwise the data are significantly different from the normal distribution.

As the sample size increases, it is assumed by central limit theorem that data follow normal distribution on approximate basis. So for very large data sets, the testing for the normality of the data become less important. Also it is generally assumed that the data follow normal distribution, when the sample size is more than 30.

W/S Test for Normality

This test is very simple test that require only the standard deviation and the range of the data.

The test statistics is defined as:

$$q = \frac{r}{s}$$

where q is the test statistics, r is the range of the data and s is the standard deviation.

Steps for applying this statistics:

- First evaluate the standard deviation of the data is square root of the variance.
- Second find the range of the data that is Highest value of the data-lowest value of the data-lowest value of the data (r=H-L).
- Evaluate the proportion $q = \frac{r}{s}$.

Now W/S test uses a critical value to test the hypothesis. If the calculated value of the test statistics falls within the critical value given in the table of W/S test. Then we accept our null hypothesis and do not reject our null hypothesis. If the calculated value falls outside the range then reject our null hypothesis.

Sometime we are interested to find out the p-value of test statistics. Then if W/S statistics i.e. $q>0.05$ then the data are not significantly different from normal distribution.

Jarque-Bera Test

A test that has similarity with the goodness of fit test. Here sample data have been tested whether skewness and kurtosis of the data matched with the normal distribution.

$$k_3 = \sum_{i=1}^{n} \frac{(y_i - \bar{y})^3}{n s^3}, k_4 = \frac{\sum_{i=1}^{n} (y_i - \bar{y})^4}{ns^4}$$

$$\text{and JB} = n\left(\frac{(k_3)^2}{6} + \frac{(k_4)^2}{24}\right).$$

Where, y_i is the ith observations of the sample data, n is the sample size, s is the standard deviation, k_3 is the skewness data and k_4 is the kurtosis of the data.

The value of the Jarque-Bera test statistic is compared with the x^2 distribution (table) with 2 degree of freedom. If the calculated value of JB > x^2 distribution (tabulated value) with 2 degree of freedom at 5% level of significance then it means that the data is significantly different from the normal distribution and our null hypothesis is rejected. On the other hand if the JB < x^2 distribution (tabulated value) with 2 degree of freedom at 5% level of significance then we do not reject the null hypothesis and the data are normally distributed.

Agostino Test

This test is considered a very powerful test to detect the departure of the data from normality. It is based on the D-statistics, which gives an upper and lower critical values.

The test statistic is defined as:

$$D=\frac{T}{\sqrt{n^3 SS_T}}$$

Where,

$$T=\sum\left(i-\frac{n+1}{2}\right)y_i$$

D is the test statistic, SS_T is the total sum of square of the observations and n is the sample size and i is the order or rank of observations (y_i).

Steps to evaluate the test statistic:

- First ordered the data in ascending order or descending order.
- Calculate the value of T by taking the difference of the ith observation rank with $\frac{n+1}{2}$ and multiply it with the y_i and take summation.
- Calculate the total sum of square i.e. subtracting each observation from the mean value of the data and square it. Now take the sum of all.
- Now put all the values in the expression of D.
- Compare the D value with the D'Agostino Normality test table.

If the value of the calculated D is less than the tabulated value then do not reject null hypothesis i.e. the data are not significant different from normal distribution. On the other hand, if the value of the test statistic is greater than the tabulated value then we reject the null hypothesis i.e. the data are not from the normal distribution.

Kolmogorov-Smirnov Test

Kolmogorov Smirnov is another test for normality. It is also known as KS test in the literature. The test statistic is defined as:

$$D_n = \max_y \left|F_{exp}(y)-F_{obs}(y)\right|.$$

Where F(y) is the cumulative distribution function of the random variable Y. For the observations y_i, $i = 1,2, \ldots, n$.

$$F_{obs}(y)=\frac{\text{no. of observation below y}}{\text{no. of observation}}$$

If we order the observation $y_{(1)} < y_{(2)} < \cdots < y_{(n)}$, then,

$$F_{obs}\left(y_{(i)}\right) = \frac{i}{n}$$

Hence first calculate the difference of observed cumulative distribution function from the observed distribution function and choose the maximum among them. This maximum value is the value of the test statistic. Now we have to compare the value of the test statistic with the tabulated value of KS table. If the value of the D_n is more than the tabulated value for given n then reject the null hypothesis otherwise do not reject the null hypothesis.

Note For large sample dataset, the critical value may not be available in the table. So there is a formula to approximate out the critical value at 5% level of significance. It is given as:

$$D_{0.05} = \frac{1.36}{\sqrt{n}}$$

Here n is the sample size.

Now the next question is, which test we should use to test for normality?

W/S Test

It is considered as the simplest test among all the test. It also has some drawback that it is not available in statistical softwares like SPSS.

Jarque Bera

This test statistic is basically used to test the normality through skewness and kurtosis and it is very effective.

D'Agostino

It is the most powerful among all the test.

Kolmogorov-Smirnov

This test is generally used in the practice and this test is not sensitive to problems in the tails. This test is only give more appropriate results when the sample size is greater than 50. This test statistic is generally available in statistical software.

As we discussed about the validity of the assumptions and how to verify them but if the data (error terms) are not from the normal distribution then what is the problem? As we already discussed that our results are only valid if our data meet some assumptions. As errors are not follow normal distribution then it may lead to serious consequences for the experimenter to make any statement about the problem.

So it is essential for a person to cross check about the normality of the data not just by graphical methods but also through the above tests.

Testing for the Equality of the Variance

The most widely test statistic that is used to test about the equality of the variance is Bartlett's test. The null hypothesis is:

$$H_0 : \sigma_1^2 = \sigma_2^2 = \cdots = \sigma_p^2$$

H_1: at least one variance is different.

The test statistic is defined as $x_{cal}^2 = 2.3026\frac{Q}{c}$

Where $Q = (np-p)\log_{10} S_p^2 - \sum_{i=1}^{p}(n-1)\log_{10} S_i^2$;

$$C = 1 + \frac{1}{3(p-1)}\left(\sum_{i=1}^{p}(n-1)^{-1} - (np-p)^{-1}\right);$$

$$S_p^2 = \frac{\sum_{i=1}^{p}(n-1)S_i^2}{(np-p)}$$

For a given sample, we can calculate the value of x_{obs}^2 We should reject the null hypothesis for large value of x_{obs}^2 that is:

$$x_{obs}^2 > x_{\alpha,p-1}^2$$

Where $x_{\alpha,p-1}^2$ is the upper α percent point of chi-square distribution with $p - 1$ degree of freedom. One can also compute the p-value to make decision about the null hypothesis.

Testing the Treatment Means using Least Square Methods

We will discuss about the testing of the treatment means. When we reject our null hypothesis of the ANOVA table for the testing of treatment means of more than two groups. The next step is to find out the pairs of the treatments that are not significantly different or to establish the ordering among the treatments means.

So if there is any significant difference between the treatment means we can take two pairs at a time to compare the treatment means between them to see which pairs are not statistically significant so that we can put them n the same group and which pairs are statistically significant so that order them according to their mean values.

The steps are given below:

- Let us suppose we have p = 4 treatments. First find the treatment mean and arrange them in ascending or descending order as:

  $$\bar{y}_3 < \bar{y}_{2.} < \cdots < \bar{y}_4$$

- In the next step, we have to examine about the treatment mean through testing by taking two pairs at a time. We know that for the testing of means of two group with unknown variance t-test is used under the assumption of equal variance.

Our hypothesis are:

$$H_o : \mu_2 = \mu_1$$
$$H_1 : \mu_2 \neq \mu_1$$

In the above hypotheses, we carries out two sided test of the hypothesis that means we are interested in to check the whether our second treatment's mean is significantly greater than or less than first treatment's mean in both directions. It would be observed from step one that we already arrange the treatment's mean in ordering so if you have an idea about the means of two treatment then we can modify of hypotheses as follows:

$$H_o : \mu_2 = \mu_1$$
$$H_1 : \mu_2 < \mu_1$$

In this case, the test statistic is called one sided test.

In general the test statistic for the two sample problem when variance is unknown is defined as:

$$\frac{(\bar{y}_1 - \bar{y}_2) - (\mu_1 - \mu_2)}{S\sqrt{\frac{1}{n_1} + \frac{1}{n_2}}} \sim t_{n1+n_2-2}$$

Where, $\bar{y}_i, i = 1,2$ are the sample mean from the first and second sample.

n_i is the sample size for the ith population,

S is the pooled variance of the both population,

μ_1 is the mean parameter of the ith population.

Here $\hat{\sigma}^2 = \frac{SS_E}{n-p}$ is an unbiased estimator of the population parameter σ^2.

Then the test statistic is given as:

$$\frac{((\bar{y}_{1.} - \bar{y}_{2.}) - (\mu_1 - \mu_2))}{\hat{\sigma}\sqrt{\frac{1}{n_1} + \frac{1}{n_2}}} \sim t_{n-p}$$

Under H_o, the test statistic is:

$$\frac{((\bar{y}_1. - \bar{y}_{2.})}{\hat{\sigma}\sqrt{\frac{1}{n_1} + \frac{1}{n_2}}}$$

Since it is a two sided test from the first hypothesis, so if there is a significant difference between the means then the statistic value must be greater than the upper 2.5% critical value of the t distribution. i.e.,

$$\frac{(\bar{y}_1.-\bar{y}_{2.})}{\hat{\sigma}\sqrt{\frac{1}{n_1}+\frac{1}{n_2}}}>t_{0.025,n-p}$$

Rearranging the statistic we get:

$$(\bar{y}_1.-\bar{y}_{2.})>t0.025,n-p\,x\,\hat{\sigma}\sqrt{\frac{1}{n_1}+\frac{1}{n_2}}.$$

The value on the right hand side of the expression is termed as the "Least significant difference" i.e. the value that the difference between the treatment means are more than this quantity then we can say that there is significant difference among the treatments.

Also if we have same sample size for all the treatments then the least significant difference will be same for all the treatments. Then the easiest way to test about the treatment mean of any pair is that if the difference between the pair of treatment mean is more than the least significant difference then we can say directly that pair of treatments are significantly different. Let us suppose 5% level of significance, we want to calculate the least significant difference between $\bar{y}_1.$ and $\bar{y}_{2.}$

The formula is,

$$t\hat{\sigma}\left(\frac{1}{n_1}+\frac{1}{n_2}\right)^{\frac{1}{2}} \quad \text{where } t=t_{0.025,n-p}$$

i.e. the value of a t distribution with n - p degree of freedom at 2.5% level of significance. If the difference of two treatment means i.e. $\bar{y}_1.-\bar{y}_{2.}$ Is less than this significant difference value then we can say that the treatment means of this pair fall into the same category or group. We can mark a symbol for them. Now we can repeat this process for other treatment pairs $\bar{y}_1.$ and $\bar{y}_{3.}$ and so on.

Randomized Block and Latin Square Design

If a large number of treatments are to be compared, then a large number of experimental units are required. This will increase the variation among the responses and CRD may not be appropriate to use. In such a case when the experimental material is not homogeneous and there are v treatments to be compared, then it may be possible to:

- Group the experimental material into blocks of sizes v units.
- Blocks are constructed such that the experimental units within a block are relatively homogeneous and resemble to each other more closely than the units in the different blocks.

- If there are b such blocks, we say that the blocks are at b levels. Similarly, if there are v treatments, we say that the treatments are at v levels. The responses from the b levels of blocks and v levels of treatments can be arranged in a two-way layout. The observed data set is arranged as follows:

		Treatments (Factor B)						Block totals
		1	2		j		v	
Blocks (Factor A)	1	y_{11}	y_{12}	...	y_{1j}	...	y_{1v}	B_1
	2	y_{21}	y_{22}	...	y_{2j}	...	y_{2v}	B_2
	⋮	⋮	⋮		⋮		⋮	⋮
	i	y_{i1}	y_{i2}	...	y_{ij}	...	y_{iv}	B_i
	⋮		⋮		⋮		⋮	⋮
	b	y_{b1}	y_{b2}	...	y_{bj}	...	y_{bv}	B_b
Treatment totals		T_1	T_2	...	T_j	...	T_v	Grand total (G)

Layout

A two-way layout is called a randomized block design (RBD) or a randomized complete block design (RCB) if, within each block, the v treatments are randomly assigned to v experimental units such that each of the v! ways of assigning the treatments to the units has the same probability of being adopted in the experiment and the assignment in different blocks are statistically independent.

The RBD utilizes the principles of design - randomization, replication and local control - in the following way:

Randomization

- Number the v treatments 1,2,...,v.
- Number the units in each block as 1, 2,...,v.
- Randomly allocate the v treatments to v experimental units in each block.

Replication

Since each treatment is appearing in each block, so every treatment will appear in all the blocks. So each treatment can be considered as if replicated the number of times as the number of blocks. Thus in RBD, the number of blocks and the number of replications are same.

Local Control

Local control is adopted in RBD in the following way:

- First form the homogeneous blocks of the experimental units.
- Then allocate each treatment randomly in each block.

The error variance now will be smaller because of homogeneous blocks and some variance will be parted away from the error variance due to the difference among the blocks.

Example: Suppose there are 7 treatments denoted as $T_1, T_2,.., T_7$ corresponding to 7 levels of a factor to be included in 4 blocks. So one possible layout of the assignment of 7 treatments to 4 different blocks in an RBD is as follows:

Block 1	T_2	T_7	T_3	T_5	T_1	T_4	T_6
Block 2	T_1	T_6	T_7	T_4	T_5	T_3	T_2
Block 3	T_7	T_5	T_1	T_6	T_4	T_2	T_3
Block 4	T_4	T_1	T_5	T_6	T_2	T_7	T_3

Analysis

Let,

y_{ij} : Individual measurements of jth treatment in i^{th} block, i = 1,2,...,b, j = 1,2,...,v.

y_{ij}'s are independently distributed following $N(\mu + \beta_i + \tau_j, \sigma^2)$

Where μ: overall mean effect

β_i : i^{th} block effect

τ_j : j^{th} treatment effect

such that $\sum_{i=1}^{b} \beta_i = 0, \sum_{j=1}^{v} \tau_j = 0$.

There are two null hypotheses to be tested.

- related to the block effects

  $H_{OB} : \beta_1 = \beta_2 = ... = \beta_b = 0.$

- related to the treatment effects

  $H_{OT} : \tau_1 = \tau_2 = ... = \tau_v = 0.$

The linear model, in this case, is a two-way model as:

$$y_{ij} = \mu + \beta_i + \tau_j + \varepsilon_{ij}, i = 1,2,..,b; j = 1,2,..,v$$

Where ε_{ij} are identically and independently distributed random errors following a normal distribution with mean 0 and variance σ^2.

The tests of hypothesis can be derived using the likelihood ratio test or the principle of least squares. The use of likelihood ratio test has already been demonstrated earlier, so we now use the principle of least squares.

Minimizing $S=\sum_{i=1}^{b}\sum_{j=1}^{v}\varepsilon_{ij}^2=\sum_{i=1}^{b}\sum_{j=1}^{v}\left(y_{ij}-\mu-\beta_i-\tau_j\right)^2$

and solving the normal equation,

$$\frac{\partial S}{\partial \mu}=0, \frac{\partial S}{\partial \beta_i}=0, \frac{\partial S}{\partial \tau_j}=0 \text{ for all } i=1,2,..,b,\ j=1,2,..,v$$

the least squares estimators are obtained as,

$$\hat{\mu}=\bar{y}_{oo},$$
$$\hat{\beta}_i=\bar{y}_{io}-\bar{y}_{oo},$$
$$\hat{\tau}_j=\bar{y}_{oj}-\bar{y}_{oo}$$

Using the fitted model (obtained after substituting the estimated values of the parameters in the model), we can write,

$$y_{ij}=\bar{y}_{oo}+\left(\bar{y}_{io}-\bar{y}_{oo}\right)+\left(\bar{y}_{oj}-\bar{y}_{oo}\right)+\left(y_{ij}-\bar{y}_{io}-\bar{y}_{oj}+\bar{y}_{oo}\right).$$

Squaring both sides and summing over i and j gives,

$$\sum_{i=1}^{b}\sum_{j=1}^{v}\left(\bar{y}_{ij}-\bar{y}_{oo}\right)^2=v\sum_{i=1}^{b}\left(\bar{y}_{io}-\bar{y}_{oo}\right)^2+b\sum_{j=1}^{v}\left(\bar{y}_{oj}-\bar{y}_{oo}\right)^2+\sum_{i=1}^{b}\sum_{j=1}^{v}\left(y_{ij}-\bar{y}_{io}-\bar{y}_{oj}+\bar{y}_{oo}\right)^2$$

Or TSS =SSBl + SSTr + SSE

with degrees of freedom partitioned as,

$$bv-1=(b-1)+(v-1)+(b-1)(v-1).$$

The reason for the number of degrees of freedom for different sums of squares is the same as in the case of CRD.

$$\text{Here TSS}=\sum_{i=1}^{b}\sum_{j=1}^{v}\left(y_{ij}-\bar{y}_{oo}\right)^2$$
$$=\sum_{i=1}^{b}\sum_{j=1}^{v}y_{ij}^2\frac{G^2}{bv}$$

$\frac{G^2}{bv}$: correction factor.

$G=\sum_{i=1}^{b}\sum_{j=1}^{v}y_{ij}$: Grand total of all the observation.

$$
\begin{aligned}
SSBI &= v\sum_{i=1}^{b}\left(\bar{y}_{io}-\bar{y}_{oo}\right)^2 \\
&= \sum_{i=1}^{b}\frac{B_i^2}{v}-\frac{G^2}{bv}
\end{aligned}
$$

$$B_i=\sum_{j=1}^{v}y_{ij}: i^{th} \text{ block total}$$

$$
\begin{aligned}
SSTr &= b\sum_{j=1}^{v}\left(\bar{y}_{oj}-\bar{y}_{oo}\right)^2 \\
&= \sum_{j=1}^{b}\frac{T_j^2}{b}-\frac{G^2}{bv}
\end{aligned}
$$

$$T_j=\sum_{i=1}^{v}y_{ij}: j^{th} \text{ treatment total}$$

$$SSE=\sum_{i=1}^{b}\sum_{j=1}^{v}\left(y_{ij}-\bar{y}_{io}-\bar{y}_{oj}+\bar{y}_{oo}\right)^2.$$

The expectations of mean squares are:

$$
\begin{aligned}
E(MSBI) &= E\left(\frac{SSBI}{b-1}\right)=\sigma^2+\frac{v}{b-1}\sum_{i=1}^{b}\beta_i^2 \\
E(MSTr) &= E\left(\frac{SSTr}{v-1}\right)=\sigma^2+\frac{v}{v-1}\sum_{j=1}^{v}\tau_j^2 \\
E(MSE) &= E\left(\frac{SSE}{(b-1)(v-1)}\right)=\sigma^2.
\end{aligned}
$$

Moreover,

$$
\begin{aligned}
&(b-1)\frac{SSBI}{\sigma^2}\sim x^2(b-1) \\
&(v-1)\frac{SSTr}{\sigma^2}\sim x^2(v-1) \\
&(b-1)(v-1)\frac{SSE}{\sigma^2}\sim x^2(b-1)(v-1)
\end{aligned}
$$

Under $H_{OB}:\beta_1=\beta_2=...=\beta_b=0,$

$$E(MSBI)=E(MSE)$$

and SSBl and SSE are independent, so,

$$F_{bl}\frac{MSBI}{MSE}\sim F\left(\left(b-1,(b-1)(v-1)\right)\right).$$

Similarly, under $H_{OT}: \tau_1 = \tau_2 = ... = \tau_v = 0$,

$$E(MSTr) = E(MSE)$$

and SSTr and SSE are independent, so

$$F_{Tr} = \frac{MSTr}{MNE} \sim F(v-1),(b-1)(v-1))$$

$$\text{Reject } H_{OB} \text{ if } F_{be} > F_{\alpha}((b-1),(b-1)(v-1)$$

$$\text{Reject } H_{OT} \text{ if } F_{Tr} > F_{\alpha}((v-1),(b-1)(v-1))$$

If H_{OB} is accepted, then it indicates that the blocking is not necessary for future experimentation.

If H_{OT} is rejected then it indicates that the treatments are different. Then the multiple comparison tests are used to divide the entire set of treatments into different subgroup such that the treatments in the same subgroup have the same treatment effect and those in the different subgroups have different treatment effects.

The analysis of variance table is as follows:

Source of variation	Degrees of freedom	Sum of squares	Mean squares	F
Blocks	b - 1	SSBl	MSBl	F_{Bl}
Treatments	v - 1	SSTr	MSTr	F_{Tr}
Errors	(b - 1)(v - 1)	SSE	MSE	
Total	bv – 1	TSS		

Latin Square Design

The treatments in the RBD are randomly assigned to b blocks such that each treatment must occur in each block rather than assigning them at random over the entire set of experimental units as in the CRD. There are only two factors – block and treatment effects – which are taken into account and the total number of experimental units needed for complete replication are bv where b and v are the numbers of blocks and treatments respectively.

If there are three factors and suppose there are b, v and k levels of each factor, then the total number of experimental units needed for a complete replication are bvk. This increases the cost of experimentation and the required number of experimental units over RBD.

In Latin square design (LSD), the experimental material is divided into rows and columns, each having the same number of experimental units which is equal to the number of treatments. The treatments are allocated to the rows and the columns such that each treatment occurs once and only once in each row and in each column.

In order to allocate the treatment to the experimental units in rows and columns, we take help from Latin squares.

Latin Square

A Latin square of order p is an arrangement of p symbols in p^2 cells arranged in p rows and p columns such that each symbol occurs once and only once in each row and in each column. For example, to write a Latin square of order 4, choose four symbols – A, B, C and D. These letters are Latin letters which are used as symbols. Write them in a way such that each of the letters out of A, B, C and D occurs once and only once in each row and each column. For example, as:

A	B	C	D
B	C	D	A
C	D	A	B
D	A	B	C

This is a Latin square.

We consider first the following example to illustrate how a Latin square is used to allocate the treatments and in getting the response.

Example:

Suppose different brands of petrol are to be compared with respect to the mileage per litre achieved in motor cars.

Important factors responsible for the variation in mileage are:

- The difference between individual cars.
- The difference in the driving habits of drivers.

We have three factors – cars, drivers and petrol brands. Suppose we have:

- 4 types of cars denoted as 1, 2, 3, 4.
- 4 drivers that are represented by a, b, c, d.
- 4 brands of petrol are indicated as A, B, C, D.

Now the complete replication will require 4×4×4=64 the number of experiments. We choose only 16 experiments. To choose such 16 experiments, we take the help of the Latin square. Suppose we choose the following Latin square:

- A B C D
- B C D A
- C D A B
- D A B C

Write them in rows and columns and choose rows for drivers, columns for cars and letter for petrol brands. Thus 16 observations are recorded as per this plan of treatment combination (as shown in the next figure) and further analysis is carried out. Since such design is based on Latin square, so it is called as a Latin square design.

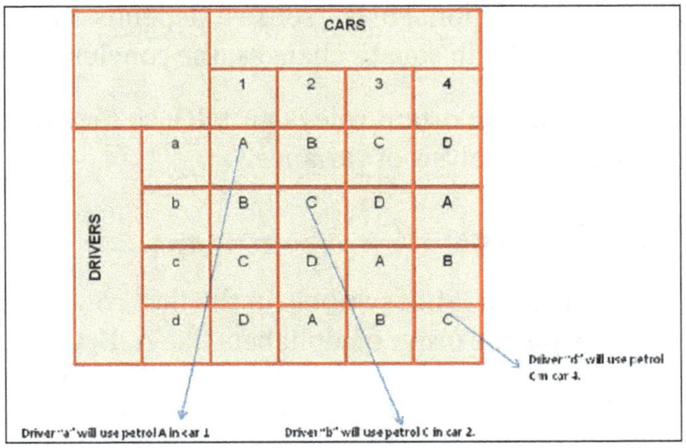

Another choice of a Latin square of order 4 is:

- C B A D
- B C D A
- A D C B
- D A B C

This will again give a design different from the previous one. The 16 observations will be recorded again but based on different treatment combinations.

Since we use only 16 out of 64 possible observations, so it is an incomplete 3-way layout in which each of the 3 factors – cars, drivers and petrol brands are at 4 levels and the observations are recorded only on 16 of the 64 possible treatment combinations.

Thus in an LSD,

- The treatments are grouped into replication in two-ways:
  - Once in rows.
  - In columns.
- Rows and columns variations are eliminated from the within treatment variation.
  - In RBD, the experimental units are divided into homogeneous blocks according to the blocking factor. Hence it eliminates the difference among blocks from the experimental error.
  - In LSD, the experimental units are grouped according to two factors. Hence two effects (like as two block effects) are removed from the experimental error.
  - So the error variance can be considerably reduced in LSD.

The LSD is an incomplete three-way layout in which each of the three factors, viz, rows, columns and treatments, is at v levels each and observations only on v^2 of the v^3 possible treatment combinations are taken. Each treatment combination contains one level of each factor.

The analysis of data in an LSD is conditional in the sense it depends on which Latin square is used for allocating the treatments. If the Latin square changes, the conclusions may also change.

We note that Latin squares play an important role is an LSD, so first we study more about these Latin squares before describing the analysis of variance.

Standard form of Latin Square

A Latin square is in the standard form if the symbols in the first row and first columns are in the natural order (Natural order means the order of alphabets like A, B, C, D,...).

Given a Latin square, it is possible to rearrange the columns so that the first row and first column remain in a natural order.

Example: Four standard forms of 4 4 × Latin square are as follows:

A B C D B A D C C D B A D C A B	A B C D B C D A C D A B D A B C	A B C D B D A C C A D B D C B A	A B C D B A D C C D A B D C B A

For each standard Latin square of order p, the p rows can be permuted in p! ways. Keeping a row fixed, vary and permute (p - 1) columns in (p - 1)! ways. So there are p!(p - 1)! different Latin squares.

For example:

Size of square	Number of Standard squares	Value of p!(1 - p)!	Total number of different squares
3 x 3	1	12	12
4 x 4	4	144	576
5 x 5	56	2880	161280
6 x 6	9408	86400	812851250

Conjugate

Two standard Latin squares are called conjugate if the rows of one are the columns of other. For example:

A B C D		A B C D
B C D A	and	B C D A
C D A B		C D A B
D A B C		D A B C

are conjugate. In fact, they are self conjugate.

A Latin square is called self conjugate if its arrangement in rows and columns are the same.

Transformation Set

A set of all Latin squares obtained from a single Latin square by permuting its rows, columns and symbols is called a transformation set.

From a Latin square of order p, p!(p - 1)! different Latin squares can be obtained by making p! permutations of columns and (p - 1)! permutations of rows which leaves the first row in place. Thus:

Number of different Latin squares of order = p!(p - 1)! X number of standard Latin squares in the set p in a transformation set.

Orthogonal Latin Square

If two Latin squares of the same order but with different symbols are such that when they are superimposed on each other, every ordered pair of symbols (different) occurs exactly once in the Latin square, then they are called orthogonal.

Graeco-Latin Square

A pair of orthogonal Latin squares, one with Latin symbols and the other with Greek symbols form a Graeco-Latin square.

For example:

A B C D $\alpha\ \beta\gamma\delta$

B A D C $\delta\ \gamma\beta\alpha$

C D A B $\beta\ \alpha\delta\gamma$

D C B A $\gamma\ \delta\alpha\ \beta$

is a Graeco-Latin square of order 4.

Graeco Latin squares design enables to consider one more factor than the factors in Latin square design. For example, in the earlier example, if there are four drivers, four cars, four petrol and each petrol has four varieties, as α, , and βγ δ , then Graeco-Latin square helps in deciding the treatment combination as follows:

	Cars				
		1	2	3	4
Drivers	a	Aα	Bβ	Cγ	Dδ
	b	Bδ	Aγ	Dβ	Cα
	c	Cβ	Dα	Aδ	Bγ
	d	Dγ	Cδ	Bα	Aβ

Now,

Aα means: Driver 'a' will use the α variant of petrol A in Car 1.

Bγ means: Driver 'c' will use the γ variant of petrol B in Car 4 and so on.

Mutually Orthogonal Latin Square

A set of Latin squares of the same order is called a set of mutually orthogonal Latin square (or a hyper Graeco-Latin square) if every pair in the set is orthogonal. The total number of mutually orthogonal Latin squares of order p is at most (p - 1).

Analysis of LSD (One Observation Per Cell)

In designing an LSD of order p,

- Choose one Latin square at random from the set of all possible Latin squares of order p.
- Select a standard Latin square from the set of all standard Latin squares with equal probability.
- Randomize all the rows and columns as follows:
    - Choose a random number, less than p, say n1 and then 2nd row is the n1th row.
    - Choose another random number less than p, say n2 and then 3rd row is the n2th n row and so on.
    - Then do the same for the column.
- For Latin squares of the order less than 5, fix the first row and then randomize rows and then randomize columns. In Latin squares of order 5 or more, need not to fix even the first row. Just randomize all rows and columns.

Example: Suppose following Latin square is chosen,

A B C D E

B C D E A

D E A B C

E A B C D

C D E A B

Now randomize rows, e.g., 3rd row becomes 5th row and 5th row becomes 3rd row. The Latin square becomes,

A B C D E

B C D E A

C D E A B

E A B C D

D E A B C

Now randomize columns, say 5th column becomes 1st column, 1st column becomes 4th column and 4th column becomes 5th column,

E B C A D

A C D B E

D A B E C

C E A D B

B D E C A

Now use this Latin square for the assignment of treatments.

y_{jik}'s are independently distributed as $N\left(\mu+\alpha_i+\beta_j+\tau_k,\sigma^2\right)$.

Linear model is:

$$y_{jik}=\mu+\alpha_i+\beta_j+\tau_k+\varepsilon_{ijk}, i=1,2,\ldots,v; j=1,2,\ldots,v; k=1,2,\ldots,v$$

Where, ε_{ijk} are random errors which are identically and independently distributed following $N\left(0,\sigma^2\right)$.

With $\sum_{i=1}^{v}\alpha_i=0, \sum_{i=1}^{v}\beta_i=0, \sum_{k=1}^{v}\tau_k=0,$

α_i Main effect of rows.

β_i Main effect of columns.

γ_k Main effect of treatments.

The null hypothesis under consideration are:

$$H_{OR}:\alpha_1=\alpha_2=\ldots.=\alpha_v=0$$
$$H_{OC}:\beta_1=\beta_2=\ldots.=\beta_v=0$$
$$H_{OT}:\tau_1=\tau_2=\ldots.=\tau_v=0$$

The analysis of variance can be developed on the same lines as earlier.

Minimizing $S=\sum_{i=1}^{v}\sum_{j=1}^{v}\sum_{k=1}^{v}\varepsilon_{ijk}^2$ with respect to μ,α_i,β_j and τ_k given the least-squares estimate as:

$$\hat{\mu}=\bar{y}_{ooo}$$
$$\hat{\alpha}=\bar{y}_{ioo}-\bar{y}_{ooo} \qquad i=1,2,\ldots,v$$
$$\hat{\beta}_j=\bar{y}_{ojo}-\bar{y}_{ooo} \qquad j=1,2,\ldots,v$$
$$\hat{\tau}_k=\bar{y}_{ook}-\bar{y}_{ooo} \qquad k=1,2,\ldots,v$$

Using the fitted model based on these estimators, the total sum of squares can be partitioned into the mutually orthogonal sum of squares SSR, SSC, SSTr and SSE as:

TSS = SSR + SSC + SSTr + SSE

Where,

TSS: Total sum of squares = $\sum_{i=1}^{v}\sum_{j=1}^{v}\sum_{k=1}^{v}\left(y_{jik}-\bar{y}_{ooo}\right)^2=\sum_{i=1}^{v}\sum_{j=1}^{v}\sum_{k=1}^{v}y_{jik}^2-\frac{G^2}{v^2}$

SSR: Sum of squares due to rows = $v\sum_{i=1}^{v}\left(y_{ioo}-\bar{y}_{ooo}\right)^2=\frac{\sum_{i=1}^{v}R_i^2}{v}-\frac{G^2}{v^2}$; where $R_i=\sum_{j=1}^{v}\sum_{k=1}^{v}y_{jik}$

SSC: Sum of squares due to column = $v\sum_{j=1}^{v}\left(y_{ojo}-\bar{y}_{ooo}\right)^2=\frac{\sum_{i=1}^{v}C_j^2}{v}-\frac{G^2}{v^2}$; where $C_j=\sum_{i=1}^{v}\sum_{k=1}^{v}y_{jik}$

SSTr : Sum of squares due to treatment = $v\sum_{k=1}^{v}\left(y_{ook}-\bar{y}_{ooo}\right)^2=\frac{\sum_{i=1}^{v}T_k^2}{v}-\frac{G^2}{v^2}$; where $T_k=\sum_{i=1}^{v}\sum_{j=1}^{v}y_{jik}$

- Degrees of freedom carried by SSR, SSC and SSTr are (v - 1) each.
- Degrees of freedom carried by TSS is v^2-1.
- Degree of freedom carried by SSE is (v - 1) (v - 2).

The expectations of mean squares are obtained as:

$$E(MSR)=E\left(\frac{SSR}{v-1}\right)=\sigma+\frac{v}{v-1}\sum_{i=1}^{v}\alpha_i^2$$
$$E(MSC)=E\left(\frac{SSC}{v-1}\right)=\sigma^2+\frac{v}{v-1}\sum_{j=1}^{v}\beta_j^2$$
$$E(MSR)=E\left(\frac{SSTr}{v-1}\right)=\sigma^2+\frac{v}{v-1}\sum_{k=1}^{v}\tau_k^2$$
$$E(MSR)=E\left(\frac{SSE}{(v-1)(v-2)}\right)=\sigma^2.$$

Thus,

$$\text{under } H_{OR}, F_R=\frac{MSR}{MSE}\sim F((v-1),(v-1)(v-2))$$
$$\text{under } H_{OC}, F_C=\frac{MSR}{MSE}\sim F((v-1),(v-1)(v-2))$$
$$\text{under } H_{OT}, F_T=\frac{MSTr}{MSE}\sim F((v-1),(v-1)(v-2))$$

Decision Rules

Reject H_{OR} at level α if $F_R>F_{1-\alpha;v(-1),(v-1)(v-2)}$

Reject H_{OC} at level α if $F_C>F_{1-\alpha;(v-1),(v-1)(v-2)}$

Reject H_{OT} at level α if $F_T>F_{1-\alpha;(v-1),(v-1)(v-2)}$.

If any null hypothesis is rejected, then use multiple comparison test.

The analysis of variance table is as follows:

Source of variation	Degrees of freedom	Sum of squares	Mean squares	F
Rows	v - 1	SSR	MSR	F_R
Columns	v - 1	SSC	MSC	F_C
Treatments	v - 1	SSTr	MSTr	F_T
Error	(v - 1)(v - 2)	SSE	MSE	
Total	$v^2 - 1$	TSS		

References

- Anova-experimental-design-analysis: iitk.ac.in, Retrieved 09, June 2020
- Completely-random-design (CRD), horsley: ndsu.edu, Retrieved 18, February 2020

6

Time Series

The series of data point which are indexed in time order is known as time series. A few of the major concepts studied in relation to time series are modeling of time series, stationary time series and non-stationary time series. The topics elaborated in this chapter will help in gaining a better perspective about these aspects of time series.

Time series is a sequence of observations with respect to study variable and time interval. It means there is a relationship between study variable and time interval.

For example: Suppose a researcher records temperature of a place on a particular day in any weather condition then recorded observation would be different in different time interval like hours, days and months. There is a relationship between temperature and time interval.

This relationship can be defined in a mathematical functional relationship. The functional relationship of time series can be described as:

$$y_t = f(t)$$

Where y_t is the value of variable under considered time t.

According to this functional relationship temperature of a place is a variable whose values are considered y_t at time t and t represents different time interval of equal length.

Classification of the Time Series

Time series can be classified in two types, first is according to dependency on time and second is according to variable numeracy.

Dependency on time can be further categorized into:

- Discrete time series.
- Continuous time series.

Variable numeracy has further two categories:

- Univariate time series.
- Multivariate time series.

We will now discuss these classification in detail:

- Discrete time series: A Time series is discrete time series when variable dependency on time is of discrete form. A variable is said to be discrete if it takes only integer values like days, months etc. For example- closing price of daily stock market.

- Continuous time series: A Time series is continuous time series when variable depends on time is of continuous nature. A variable is said to be continuous if it takes values in continuous form like in second, minutes etc. For example- hourly reading of temperature.
- Univariate time series: A time series is univariate time series when there is only one type of study variable or one characteristics under consideration. For example: share price of one company.
- Multivariate time series: A Time series is multivariate time series when there is more than one type of study variable that is of interest in time series. For example- share price of one company, share price of other company.
- Components of time series: Study variable take different values in a particular time series. This difference in value is not due to only one component like time but affected by more than one component or multiple components. These multiple components pull up and down the values of the characteristics under study. For example: the crop yield depends on weather condition, seed quality, land quality, fertilizer etc. Here crop yield is study variable and weather condition, seed quality, land quality are other multiple components that are also responsible for the yield of the crop.

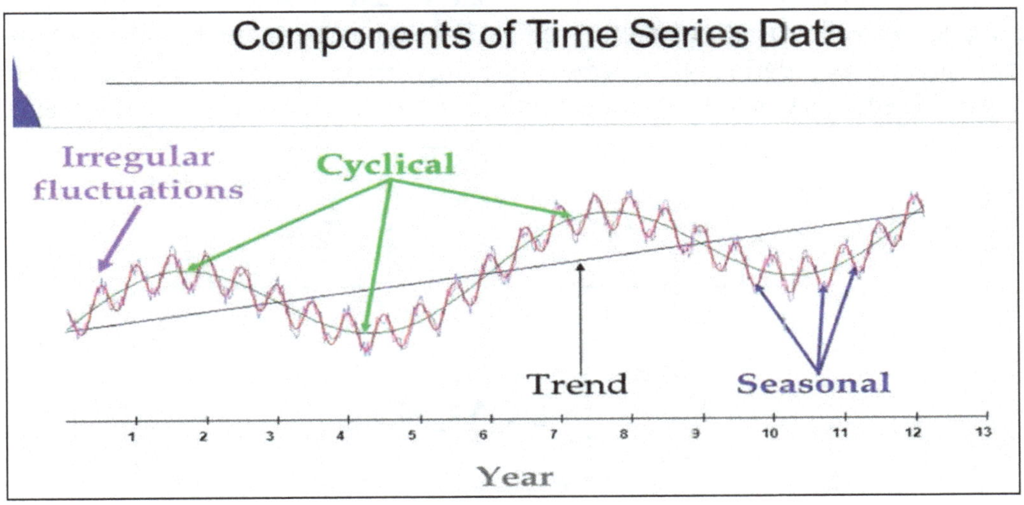

These multiple components are classified into four categories. Such as:

- Secular Trend (long term movement).
- Seasonal variation (short term movement).
- Cyclic variation.
- Random or Irregular Movements.

These four categories are known as components of time series.

Secular Trend

Secular trend or generally trend is a long term movement. In general, Trend shows tendency of the time series data in long term period. This tendency may be downward and upward. It should be kept in mind that this tendency of upward and downward movement should not remain same throughout the time span. It is also possible that it can be observed in different time section or time span. However one can say that the overall tendency of the series seem like moving upward or downward but in actual there are many factor that persist for a long period. Tendency depends on the effects of the factor. If effect of factors on time series data is of increasing then it shows upward tendency, if is on decreasing then it shows downward tendency.

Trend consists of Long term effects. Long term cannot be defined exactly because in some cases 2 year time may be enough and for some cases it is not enough.

For some cases 5 minutes is a very long period, for other case it may not be appropriate. The term of long or short period of time depends on the nature or objective of the data. In some cases, a period of hours can be considered as very large while in other cases even the period of years is not sufficient to be named as long time. For example if data of agriculture production of 24 month shows increment in the growth of the crop then it is not considered as secular trend for 2 years, the count of bacterial population of a culture in every five minutes, if this tendency of increment persist for a week then it is a secular trend. One important thing is that the values for short period (may be 2-3 years in some cases) are mainly affected by cyclic variations. Hence the values are not able to reveal true trend. In order to solve this problem, one should consider the multiple cycle periods to see the impact of cyclic variations on the values. For some cases, due to nature of the data like increment of bacterial population persists due to strong germicide in an hour then reading in every five minutes or in seconds would lead to a general pattern that can be termed as a secular trend.

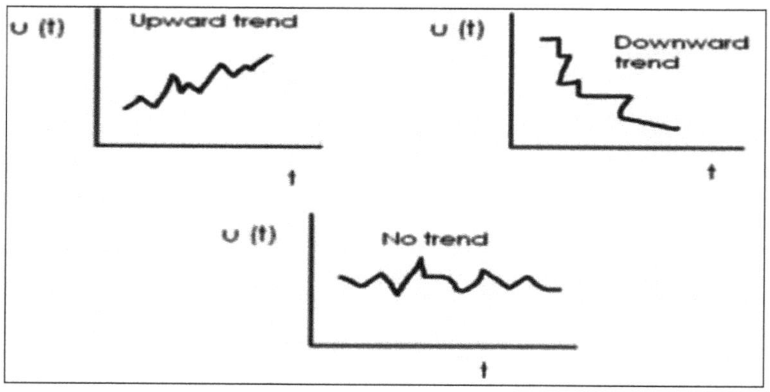

This tendency of the data may be linear and nonlinear trend. It depends on the nature of the data. If the time series values are plotted on the graph with more concentration around a straight line then trend pattern of time series values are said to be linear on the other hand if the time series values are not close to straight line then it is called non-linear trend like increase in the price of gold, crude oil in the international market is not constant and varies with different phases of time. Hence the trend of time series will help you to get a general idea about the pattern of the behavior of the data under consideration. This will help in forecasting and future planning of different sectors like business, weather and economy based on previous values. With the use of trend analysis one can compare two or more time series over different time periods and draw important conclusions based on them.

Seasonal Variation

Seasonal variation is a short term movement in a time series. Short term movement means observational data collects in less than one year of time period i.e. data collected monthly, quarterly, weekly, daily, hourly etc. If collected data are annual then there is no seasonal variation. As the name suggests, seasonal variation means variation that occur due to change in season and natural forces in any study variable. For example prices, production and consumption of any commodities, sales and profit in any departmental store etc.

Some variations in data are due to natural forces like seasons. For example: sale of umbrella increases in rainy season, consumption of ice-cream increases in summer, price of woolen increases in winter. Some variation are due to man-made convention such as sale of jewellery and price of clothes go up due to marriage season and in festival like Christmas, New Year and Easter.

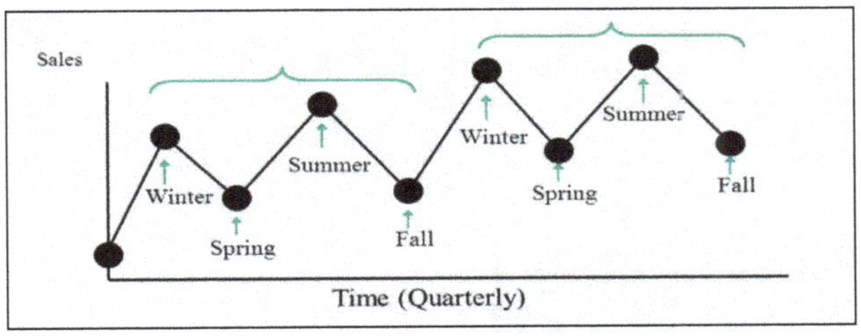

Sales of clothes changing in different seasons.

Cyclic Variation

Cyclic variation is a repeated pattern in a time period of a time series, this time period will be more than one year. The data of sale of ice cream for few years shows that the sale of ice-cream is high in summer and low in winter. Therefore this data shows some repeated pattern in time series data, this pattern is known as cyclic variation .

Generally cyclic variation shows in business cycle. Business cycle have four phases. Hence this is referred as four phase cycle such as prosperity, recession, depression and recovery. Each phase in business cycle persists for a very long term variation approximately 5 to 7 years. On the other side, cyclic variation may be short term and long term movement that depend on time period but time period must be greater than one year.

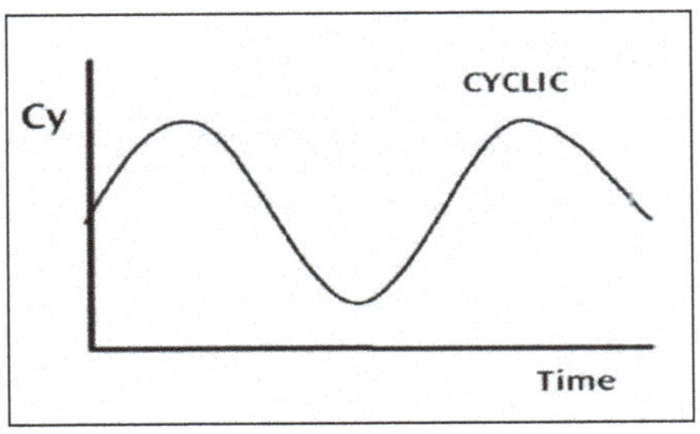

Irregular Component

Irregular component is also known as random or residual component. Every time series must contain this fluctuation and this fluctuation cannot be removed from the data. Irregular component can be found by removing all three components, so that irregular component is not affected by other three components of time series.

Irregular component is beyond the control of human being and considered as unpredictable. This type of fluctuation may be seen due to effect of factors like earthquake, wars, floods, famines, political unrest revolution, epidemics etc.

Irregular variation are unpredictable due to its accidental variation.

For example- rise in prices of gasoline due to strike of workers etc.

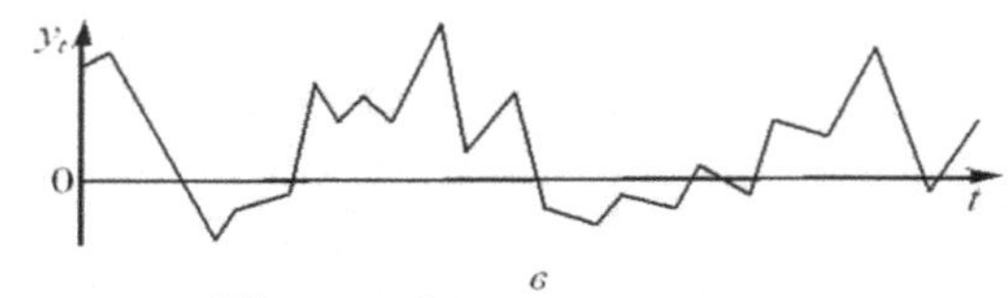

The random component

Modeling of Time Series

In analysis of time series, there are types of problems like to identify the components and their effect on the data sets, to isolate, analyze and measure them independently. In literature, there are some mathematical model to decompose the time series into its components.

Additive model: Additive model of a time series can be expressed as:

$$y_t = T_t + S_t + C_t + R_t$$

where y_t is time series value at time t.

T_t is trend value at time t.

S_t is seasonality value at time t.

C_t is cyclic value at time t.

R_t is random or irregular value at time t.

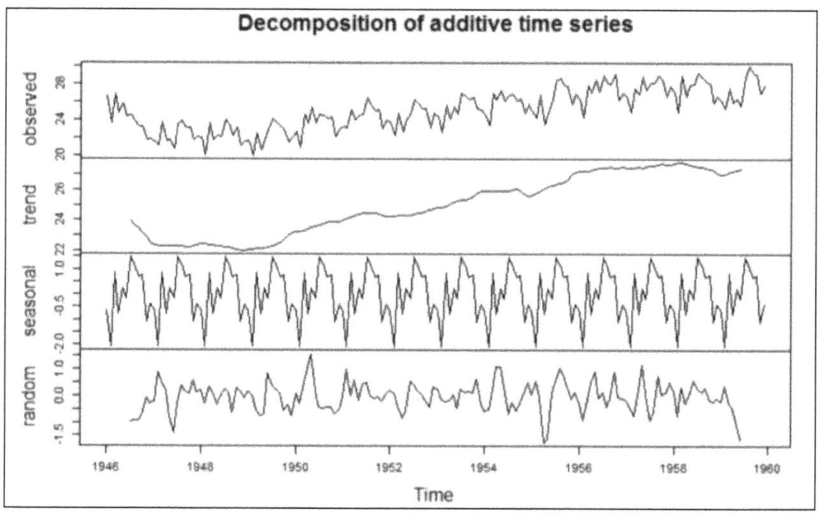

All four component of time series in additive model operate independently to each other. Thus, in additive model, there is no effect of one component to other component. Additive model can be used to measure one or more component by elimination i.e. by subtraction.

For example: The time series decomposition of additive model with visualization as in the above image.

Multiplicative Model

If all component of time series operate proportionally to the general level of the series, multiplicative model is appropriate. Multiplicative model can be expressed as:

$$y_t = T_t X S_t X C_t X R_t$$

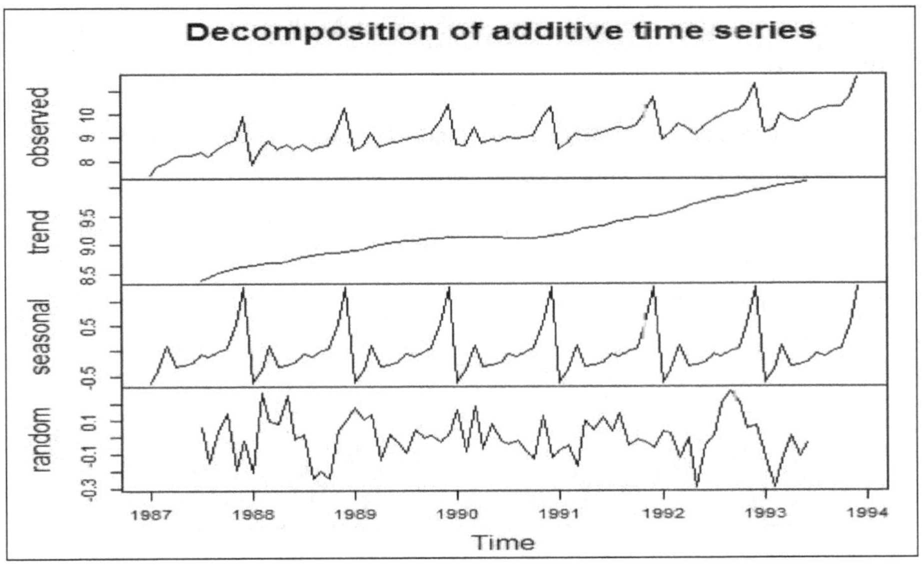

Where,

y_t is the value of time series at time t.

T_t, S_t, C_t and R_t is the value of trend, seasonal, cyclic and irregular component respectively. Multiplicative decomposition of time series is same as the additive decomposition of logarithmic values of the original time series that is:

$$\log y_t = \log T_t + \log S_t + \log C_t + \log R_t$$

For example: time series transform Multiplicative to Additive decomposition by using logarithmic with visualization is given in figure.

It means that when the observations are related with each other in proportionate manner then by taking log of the observation we can convert the multiplicative model into additive model as it is easier to detect the components in additive models than multiplicative model. By taking log of observations one can smooth the curve. In practical cases, it is most widely used concept to take log of observations to determine the model that fits in an appropriate manner on the observations.

Most of time series related to business and economic confirms to multiplicative model. Multiplicative model are used in measure of one or more components i.e. by division. For example if trend value is known then isolate them:

$$S_t X C_t X R_t = \frac{y_t}{T_t}$$

If data is annual then seasonal components is not there i.e.

$$C_t X R_t = \frac{y_t}{T_t}$$

For example- Assume birth time series decomposition of Multiplicative model with visualization as in the following image:

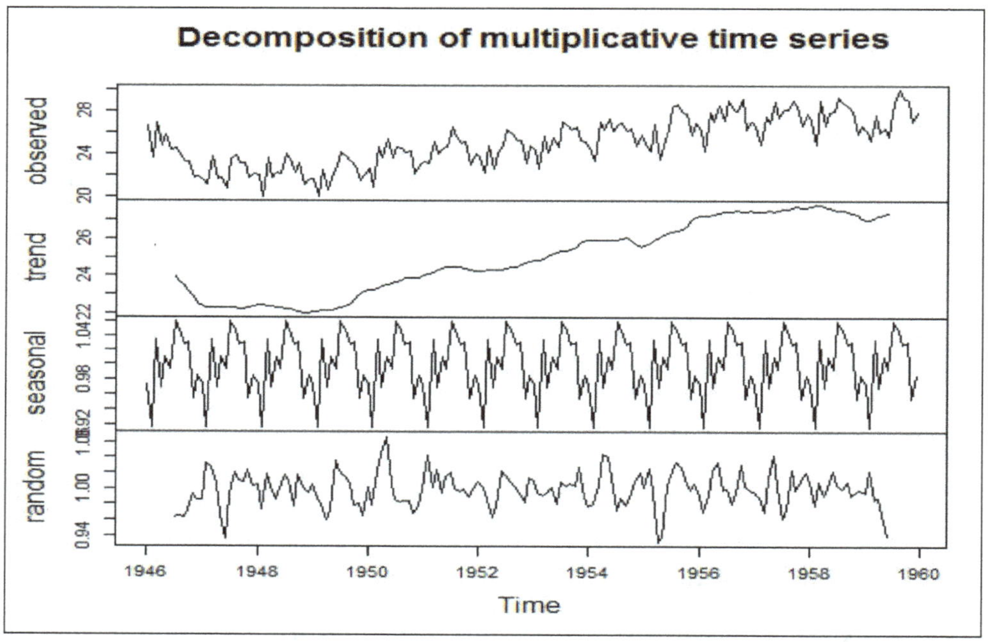

Hence from figures, one can understand the mathematical model of time series components. One can choose an appropriate model based on the relationship among components in the data values.

Mixed Model

Some time series data shows different type of models under different type of assumption. This different type of model is known as combination of additive and multiplicative model or mixed model of time series. Some types of different model are:

$$y_t = T_t X C_t + S_t X R_t$$
$$y_t = S_t X C_t X R_t + T_t$$
$$y_t = T_t + S_t + C_t X R_t$$

Where,

y_t is the value of time series at time t.

T_t, S_t, C_t and R_t is the value of trend, seasonal, cyclic and irregular component.

Utility of Time Series

Utility of time series is given as:

- It helps in understanding past behavior of the time series: To understand the past behavior of any time series, to analyze variation between data sets over a period of time.
- It helps in future prediction: By observing past behavior of time series data we may see how observation are varying. According to behavior of data one can use this information for forecasting. We know that time series depend on past data. For better prediction of values one require long period data.
- It helps in understand current situation of the problem: The actual performance can be analyzed by taking difference between expected performance and cause of variation of the data.
- It facilitates comparisons between different time series.

Requirements of Time Series

Time series have following requirements:

- Data must consist of a homogenous set of values.
- The data should be available for sufficiently long period.
- Time gap between various values must, as far as possible be equal.
- The gaps, if any, in data should be made up by interpolation.

Application of Time Series

Time series analysis has wide application area. Most important area of time series analysis is economic and business. Some other application areas are: weather forecasting, pattern recognition, earthquake prediction, mathematical science and mathematical finance. Time series analysis is mostly used in forecasting or predication.

Stationary Time Series

A stationary time series is one whose properties do not depend on the time at which the series is observed. Thus, time series with trends, or with seasonality, are not stationary — the trend and seasonality will affect the value of the time series at different times. On the other hand, a white noise series is stationary — it does not matter when you observe it, it should look much the same at any point in time.

Some cases can be confusing — a time series with cyclic behaviour (but with no trend or seasonality) is stationary. This is because the cycles are not of a fixed length, so before we observe the series we cannot be sure where the peaks and troughs of the cycles will be.

In general, a stationary time series will have no predictable patterns in the long-term. Time plots will show the series to be roughly horizontal (although some cyclic behaviour is possible), with constant variance.

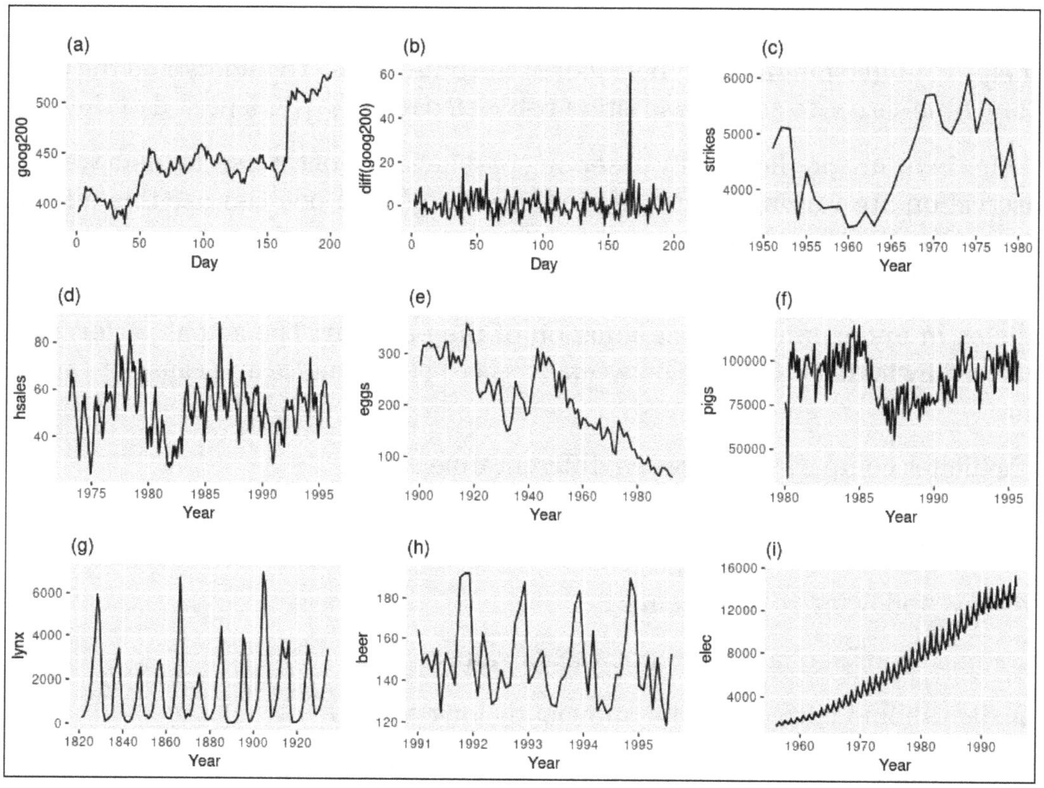

Which of these series are stationary? (a) Google stock price for 200 consecutive days; (b) Daily change in the Google stock price for 200 consecutive days; (c) Annual number of strikes in the US; (d) Monthly sales of new one-family houses sold in the US; (e) Annual price of a dozen eggs in the US (constant dollars); (f) Monthly total of pigs slaughtered in Victoria, Australia; (g) Annual total of lynx trapped in the McKenzie River district of north-west Canada; (h) Monthly Australian beer production; (i) Monthly Australian electricity production.

Consider the nine series plotted in figure. Which of these do you think are stationary?

Obvious seasonality rules out series (d), (h) and (i). Trends and changing levels rules out series (a), (c), (e), (f) and (i). Increasing variance also rules out (i). That leaves only (b) and (g) as stationary series.

At first glance, the strong cycles in series (g) might appear to make it non-stationary. But these cycles are aperiodic — they are caused when the lynx population becomes too large for the available feed, so that they stop breeding and the population falls to low numbers, then the regeneration of their food sources allows the population to grow again, and so on. In the long-term, the timing of these cycles is not predictable. Hence the series is stationary.

Differencing

In figure, note that the Google stock price was non-stationary in panel (a), but the daily changes were stationary in panel (b). This shows one way to make a non-stationary time series stationary — compute the differences between consecutive observations. This is known as differencing.

Transformations such as logarithms can help to stabilise the variance of a time series. Differencing can help stabilise the mean of a time series by removing changes in the level of a time series, and therefore eliminating (or reducing) trend and seasonality.

As well as looking at the time plot of the data, the ACF plot is also useful for identifying non-stationary time series. For a stationary time series, the ACF will drop to zero relatively quickly, while the ACF of non-stationary data decreases slowly. Also, for non-stationary data, the value of r_1 is often large and positive.

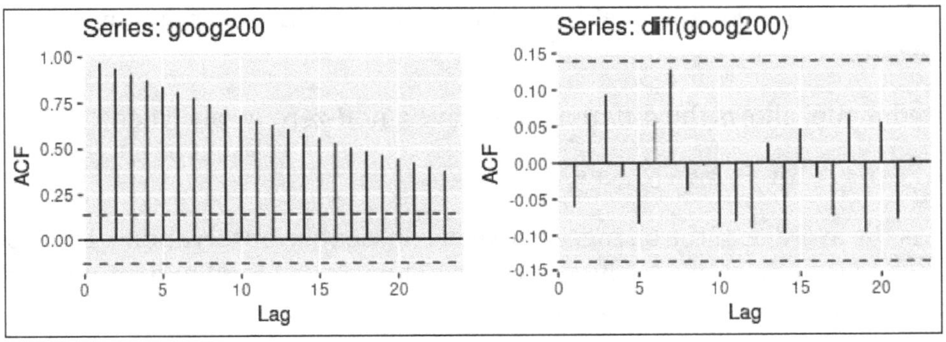

The ACF of the Google stock price (left) and of the daily changes in Google stock price (right).

```
Box.test(diff(goog200), lag=10, type="Ljung-Box")
#>
#>  Box-Ljung test
#>
#> data:  diff(goog200)
#> X-squared = 11, df = 10, p-value = 0.4
```

The ACF of the differenced Google stock price looks just like that of a white noise series. There are no autocorrelations lying outside the 95% limits, and the Ljung-Box Q^* statistic has a *p*-value of 0.355 (for h=10). This suggests that the *daily change* in the Google stock price is essentially a random amount which is uncorrelated with that of previous days.

Random Walk Model

The differenced series is the *change* between consecutive observations in the original series, and can be written as:

$$y'_t = y_t - y_{t-1}.$$

The differenced series will have only T−1 values, since it is not possible to calculate a difference y'_1 for the first observation.

When the differenced series is white noise, the model for the original series can be written as:

$$y_t - y_{t-1} = \varepsilon_t,$$

where ε_t denotes white noise. Rearranging this leads to the "random walk" model:

$$y_t = y_{t-1} + \varepsilon_t,$$

Random walk models are widely used for non-stationary data, particularly financial and economic data. Random walks typically have:

- Long periods of apparent trends up or down.
- Sudden and unpredictable changes in direction.

The forecasts from a random walk model are equal to the last observation, as future movements are unpredictable, and are equally likely to be up or down. Thus, the random walk model underpins naïve forecasts.

A closely related model allows the differences to have a non-zero mean. Then,

$$y_t - y_{t-1} = c + \varepsilon_t \text{ or } y_t = c + y_{t-1} + \varepsilon_t.$$

The value of c is the average of the changes between consecutive observations. If c is positive, then the average change is an increase in the value of y_t. Thus, y_t will tend to drift upwards. However, if c is negative, y_t will tend to drift downwards.

This is the model behind the drift method.

Second-Order Differencing

Occasionally the differenced data will not appear to be stationary and it may be necessary to difference the data a second time to obtain a stationary series:

$$\begin{aligned} y''_t &= y'_t - y'_{t-1} \\ &= (y_t - y_{t-1}) - (y_{t-1} - y_{t-2}) \\ &= y_t - 2y_{t-1} + y_{t-2}. \end{aligned}$$

In this case, y''_t will have T−2 values. Then, we would model the "change in the changes" of the original data. In practice, it is almost never necessary to go beyond second-order differences.

Seasonal Differencing

A seasonal difference is the difference between an observation and the previous observation from the same season. So,

$$y'_t = y_t - y_{t-m},$$

where m = the number of seasons. These are also called "lag-mm differences," as we subtract the observation after a lag of mm periods.

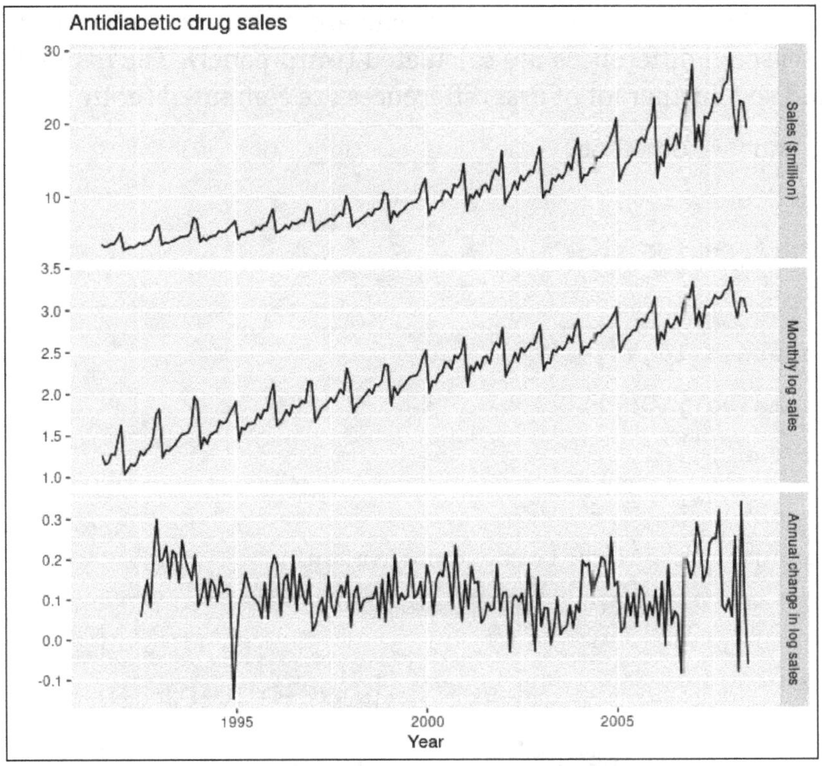

Logs and seasonal differences of the A10 (antidiabetic) sales data. The logarithms stabilise the variance, while the seasonal differences remove the seasonality and trend.

If seasonally differenced data appear to be white noise, then an appropriate model for the original data is:

$$y_t = y_{t-m} + \varepsilon_t.$$

Forecasts from this model are equal to the last observation from the relevant season. That is, this model gives seasonal naïve forecasts.

The bottom panel in figure shows the seasonal differences of the logarithm of the monthly scripts for A10 (antidiabetic) drugs sold in Australia. The transformation and differencing have made the series look relatively stationary.

```
cbind("Sales ($million)" = a10,
"Monthly log sales" = log(a10),
"Annual change in log sales" = diff(log(a10),12)) %>%
autoplot(facets=TRUE) +
xlab("Year") + ylab("") +
ggtitle("Antidiabetic drug sales")
```

To distinguish seasonal differences from ordinary differences, we sometimes refer to ordinary differences as "first differences," meaning differences at lag 1.

Sometimes it is necessary to take both a seasonal difference and a first difference to obtain stationary data, as is shown in figure. Here, the data are first transformed using logarithms (second panel), then seasonal differences are calculated (third panel). The data still seem somewhat non-stationary, and so a further lot of first differences are computed (bottom panel).

```
cbind("Billion kWh" = usmelec,
"Logs" = log(usmelec),
"Seasonally\n differenced logs" =
diff(log(usmelec),12),
"Doubly\n differenced logs" =
diff(diff(log(usmelec),12),1)) %>%
autoplot(facets=TRUE) +
xlab("Year") + ylab("") +
ggtitle("Monthly US net electricity generation")
```

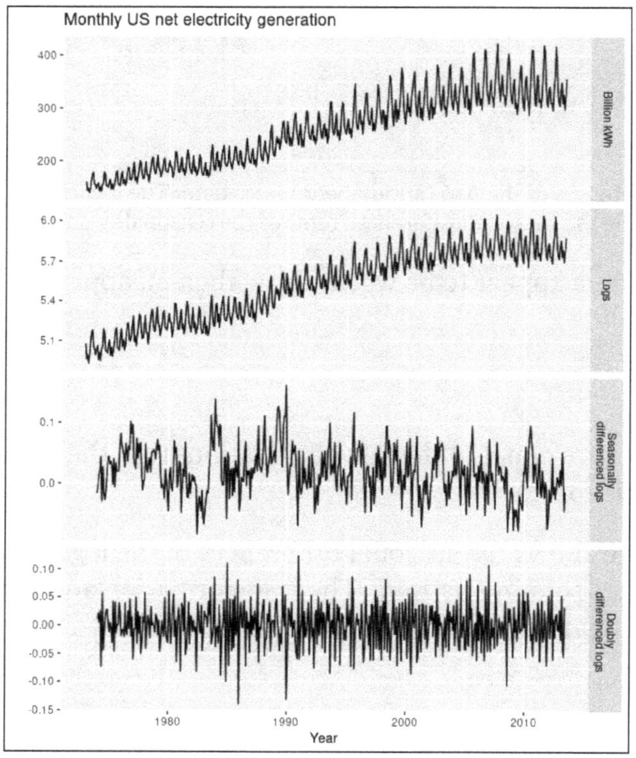

Top panel: US net electricity generation (billion kWh). Other panels show the same data after transforming and differencing.

There is a degree of subjectivity in selecting which differences to apply. The seasonally differenced data in figure do not show substantially different behaviour from the seasonally differenced data in figure. In the latter case, we could have decided to stop with the seasonally differenced data, and not done an extra round of differencing. In the former case, we could have decided that the data were not sufficiently stationary and taken an extra round of differencing. Some formal tests for differencing are discussed below, but there are always some choices to be made in the modelling process, and different analysts may make different choices.

If $y'_t = y_t - y_{t-m}$ denotes a seasonally differenced series, then the twice-differenced series is:

$$\begin{aligned} y''_t &= y'_t - y'_{t-1} \\ &= (y_t - y_{t-m}) - (y_{t-1} - y_{t-m-1}) \\ &= y_t - y_{t-1} - y_{t-m} + y_{t-m-1} \end{aligned}$$

When both seasonal and first differences are applied, it makes no difference which is done first—the result will be the same. However, if the data have a strong seasonal pattern, we recommend that seasonal differencing be done first, because the resulting series will sometimes be stationary and there will be no need for a further first difference. If first differencing is done first, there will still be seasonality present.

It is important that if differencing is used, the differences are interpretable. First differences are the change between one observation and the next. Seasonal differences are the change between one year to the next. Other lags are unlikely to make much interpretable sense and should be avoided.

Non-Stationary Time Series

If the graphical representation of the recorded data shows upward or downward trend, cycles, random walks with/without combination of trend, cycle and random walks, then time series is non-stationary data and have non-stationarity condition.

In the figure below, one can see the different components and their combinations that exits among different time series through different colors for better understanding.

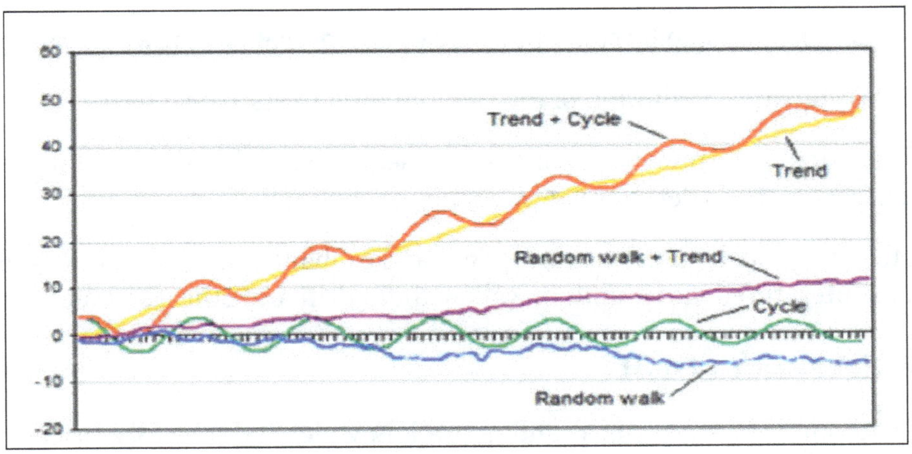

Non-Stationary data possess non-stability into the mean and variance in long run. This means that variance goes to infinity as time goes infinite and mean value never become stable on one value as time passes.

Non-Stationary data can't be predicted, modeled and forecasted due to presence of non-stability factors that lead to variation in the values of statistical properties like mean, variance over time.

To get consistent and reliable outcome of non-stationary data, there is a need to transform non-stationary data into stationary data. Under stationarity condition forecasted value will be consistent and reliable.

Types of Non-Stationary Processes

Before discussing the transformation of the non-stationary time series into stationary time series, different types of non-stationary process will be discussed. These non-stationary process may be helpful in the transformation of the non-stationary process.

Some well known non-stationary processes are:

- Random walk without drift.
- Random walk with drift.
- Deterministic trend.

Random Walk

In simple term random walk is a sequence of random changes, either in the variable or process, subject to condition. Random walk is also defined as a path taken by a point and quantity that moves in steps and direction of the each step will be decided randomly.

Suppose you want to predict chances of accident of a drunk who is walking on road. One truck is coming towards him and he is walking randomly. He is in the middle of the road now and truck is coming towards him. He takes steps randomly left, right, forward or backward from the current position. You can't predict if that person will meet with an accident because he is taking steps randomly in random direction.

Drunk person taking moves step by step randomly so this process is known as random walk.

Now let us discuss statistical definition of the random walk in statistical terms.

"Random walk is a stochastic process consisting of a sequence of changes in stepwise manner each of which step characteristic is determined by chance or randomly".

Random walk theory is mostly used in stock prices and exchange rates. Random walk is discussed in two ways in the literature, one is random walk without drift and other is random walk with drift.

Random Walk Without Drift

Random walk without drift; simply is a random walk in which today's stock price will be equal to yesterday's stock price plus a random stock. This is a nonmean reverting process which means that it can move towards positive and negative side and variance changes as time changes in long run, so that it can't be predicted.

Mathematical expression for random walk without drift is:

$$P_t = P_{t-1} + u_t$$

where P_t is the price of the stock at time t;

P_{t-1} is the price of the stock at time t– 1;

u_t is the variable that take positive or negative value.

Here the price of the stock at time t will depend on the price of the stock at time t-1 and the random movement of the stock price at time t.

Random Walk with Drift

Random walk with drift, tomorrow's value will be equal to the today's value plus drift plus random variable. Drift is the constant and average value of measuring changes. It can be used to measure trend, cycle. It is not accurately predictable because there is no constant mean and variance.

Mathematical expression for the random walk with drift:

$$P_t = \beta + P_{t-1} + u_t$$

where *Pt* is the price of the stock at time t;

β is the constant parameter;

P_{t-1} is the price of the stock at time t-1;

u_t is the variable that take positive or negative value.

Here the price of the stock at time t depends on the average value of the stock plus price of the stock at time t-1 and random movement at time t.

References

- Statistical-Applications-in-Environmental-Sciences: inflibnet.ac.in. Retrieved 10, February 2020
- Stationarity-and-differencing : otexts.com, Retrieved 21, April 2020

Permissions

All chapters in this book are published with permission under the Creative Commons Attribution Share Alike License or equivalent. Every chapter published in this book has been scrutinized by our experts. Their significance has been extensively debated. The topics covered herein carry significant information for a comprehensive understanding. They may even be implemented as practical applications or may be referred to as a beginning point for further studies.

We would like to thank the editorial team for lending their expertise to make the book truly unique. They have played a crucial role in the development of this book. Without their invaluable contributions this book wouldn't have been possible. They have made vital efforts to compile up to date information on the varied aspects of this subject to make this book a valuable addition to the collection of many professionals and students.

This book was conceptualized with the vision of imparting up-to-date and integrated information in this field. To ensure the same, a matchless editorial board was set up. Every individual on the board went through rigorous rounds of assessment to prove their worth. After which they invested a large part of their time researching and compiling the most relevant data for our readers.

The editorial board has been involved in producing this book since its inception. They have spent rigorous hours researching and exploring the diverse topics which have resulted in the successful publishing of this book. They have passed on their knowledge of decades through this book. To expedite this challenging task, the publisher supported the team at every step. A small team of assistant editors was also appointed to further simplify the editing procedure and attain best results for the readers.

Apart from the editorial board, the designing team has also invested a significant amount of their time in understanding the subject and creating the most relevant covers. They scrutinized every image to scout for the most suitable representation of the subject and create an appropriate cover for the book.

The publishing team has been an ardent support to the editorial, designing and production team. Their endless efforts to recruit the best for this project, has resulted in the accomplishment of this book. They are a veteran in the field of academics and their pool of knowledge is as vast as their experience in printing. Their expertise and guidance has proved useful at every step. Their uncompromising quality standards have made this book an exceptional effort. Their encouragement from time to time has been an inspiration for everyone.

The publisher and the editorial board hope that this book will prove to be a valuable piece of knowledge for students, practitioners and scholars across the globe.

Index